ENERGY FROM SOLID WASTES

POLLUTION ENGINEERING AND TECHNOLOGY

A Series of Reference Books and Textbooks

EDITORS

RICHARD A. YOUNG | PAUL N. CHEREMISINOFF

Editor, Pollution Engineering
Technical Publishing Company
Barrington, Illinois

Associate Professor of
Environmental Engineering
New Jersey Institute of
Technology
Newark, New Jersey

1. Energy from Solid Wastes, *Paul N. Cheremisinoff and Angelo C. Morresi*

Additional Volumes in Preparation

ENERGY FROM SOLID WASTES

Paul N. Cheremisinoff

Angelo C. Morresi

MARCEL DEKKER, INC.   New York and Basel

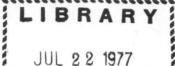
LIBRARY OF CONGRESS CATALOG CARD NUMBER: 76-1799

ISBN: 0-8247-6454-4

Current printing (last digit):
10 9 8 7 6 5 4 3 2 1

PRINTED IN THE UNITED STATES OF AMERICA

PREFACE

Modern society today is the product of constantly advancing technology. Technological progress is responsible for our high overall living standard and also for pollution and waste. Fortunately, technology can also be used to solve these problems.

Over the past decade man has begun to consider the importance of his environment and his impact on it. Concerns for the environment are not limited to detrimental effects of pollution but also include recovery and utilization of resources now recognized as finite, namely fuels and energy. Solid wastes problems began with Adam throwing away the first apple core. Energy problems became prominent with oil embargos and rising fuel prices in 1974.

Tremendous savings can be realized by many industries, institutions and governments through the potential of converting their own solid wastes into energy. Methods are available or being developed that are highly engineered and environmentally clean. These recovery possibilities can open a new era in converting waste into energy and enable the originators of the waste to recycle so-called wastes into heat and power, replacing dwindling supplies of natural gas and increasingly expensive oil and coal.

The energy locked up in trash in the United States alone amounts to over 900 trillion BTU annually. Some of this can be released by merely lighting with a match the nearly 300 million

tons of garbage we produce. Other wastes and methods appear

technologically attractive and require highly engineered systems,

sophistication and added ingenuity. However, with oil at 30¢/gallon

this waste is worth about $25/ton, and it's bad economics to pay

someone to haul it away. There have been estimates that industry

depends on natural gas for 51 percent of its energy at a time when

natural gas authorities tell us such users may have supply cut

90-95 percent by 1980. What will this mean in terms of jobs and

income over the next five years?

The purpose of this book is to furnish engineers, managers,

and students with the possibilities of wastes as an energy source.

The benefits are mainly two-fold: (1) a potentially significant

contribution to the total energy supply, and (2) relief from the

many environmental problems associated with wastes' disposal.

Paul N. Cheremisinoff
Angelo C. Morresi

CONTENTS

ENERGY FROM SOLID WASTES

SOLID WASTE AS A POTENTIAL FUEL

Introduction - The Fossil Fuels

The expression "energy crisis" has evolved to mean much more than a temporary fuel shortage based on the political differences of the western world and oil-producing Arab states. Today, our shortage of fuel supplies is no longer believed to be merely a conspiracy contrived by the major oil companies, but an actual fact of life. The fossil fuel supplies of our planet are limited. Exploration for new oil supplies continues at record levels, both in monetary and drilling terms, yet without much foresight. New petroleum discoveries of even fantastic proportions will only shift the world into a false sense of security; for oil reserves of even 50 years would put the problem on our children's shoulders and alternate energy means would then become a necessity. For at some point in time these fuels will probably be used up or diminished to a point of economical or environmental unattractiveness.

Another major factor limiting the production and use of
fossil fuels is environmental damage. In the effort to obtain these
fossil fuels irreparable harm is, and can be done, to the environment
as a result of strip mining; oil spills from tankers and off-shore
drilling platforms; pollutant emissions from loading and unloading
tankers and storage vessels; pipelines; air and water effluents
from the refinery processes; coal gasification; coal liquefaction;
and pollutant emissions effected by the burning of these fuels
for power generation. Thus, fossil fuels are rapidly becoming
a precious commodity which has been reflected in their increasing
prices. Within the past two years the price of gasoline at the
pump has nearly doubled and coal costs have increased over 50 percent. Natural gas prices have also doubled and the price of
petroleum products in general have gone the way of gold - up.
The future expects further increases as a result of governmental
intervention into prices including fuel allotments, controls,
decontrols and a committment by Washington to get our country
independent of foreign oil.

Figures 1 through 3 estimate the available fossil fuels.
Fuel oils and natural gas reserves can be used up in our lifetimes
while the possibilities for coal seem more attractive. Coal is
present through the United States in relatively infinite quantities;
however, environmental technology has not progressed to the level
where utilizing these coal reserves would be beneficial to the
populus. Figure 4 exemplifies America's dependency on the fossil
fuels for its energy sources and nearly 75 percent of its require-

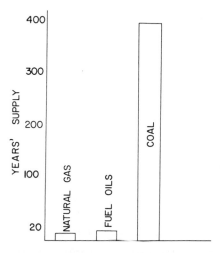

FIG. 1. Fossil fuel reserves.

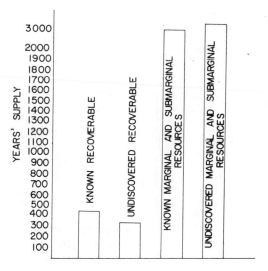

FIG. 2.   Estimates of U. S. coal reserves.[1]

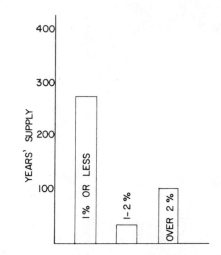

FIG. 3.   Coal reserves distribution according to sulfur content.

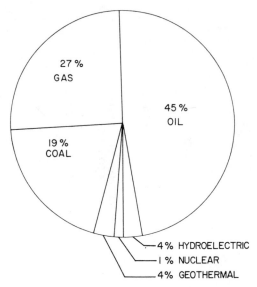

FIG. 4.   U. S. energy consumption sources.

ments are based on natural gas and oil. Figure 1 shows that these
fuels have reserves which will last less than twenty years.

Thus, the energy shortage has many facets and some
alternatives that may be considered are:

. For the short term, an increased dependence on coal,
and nuclear power.

. The conservation of the more precious fossil fuels;
natural gas and oil.

. The research and development of technologies for
alternate energy sources.

. The making available of energy sources such as solid
wastes, by removing archaic legal procedures that tend to increase
costs and create problems.

The Potential of Municipal Solid Wastes

The most common material found in municipal solid wastes
is paper in forms such as newspaper, cardboard, magazines, and
brown paper. These different forms of paper make up more than half
the weight of a typical municipal refuse. An analysis, made by the
Environmental Protection Agency to determine the composition of
typical municipal refuse, found that 53 percent was paper, 8 percent
was organic scraps, 8 percent glass, 7 percent metals, and 24 percent
rags and various household items. This pound of refuse includes
moisture, ash, metals, and combustible material and would have a
heating value of 5,300 Btu per pound or 10.5 million Btu per ton.

One desirable quality in solid wastes is the low sulfur
content of the materials, usually less than 0.12 percent by weight.
Therefore, refuse used as a fuel in steam plants would be able to
meet federal low sulfur fuel standards. Typical standards allow
approximately 0.5 pounds per million Btu sulfur, whereas solid
waste contains 0.23 pounds per million Btu or about half of the
federal requirements.

A problem in using solid wastes for fuel is its varying
composition. The components of solid wastes also vary regionally
and seasonally. When solid waste is utilized as a fuel the 53 percent
paper content will effect 70 percent of the heat output. Encouraging
the burning of solid wastes is the fact that upon combustion, paper
will release up to one-third of the heat that was needed to process
it and upon disposal in a landfill this heat is not utilized.

The burning of refuse as a fuel source as opposed to
incineration necessitates priority be given to the refuse's heating
value along with a mass and volume reduction. Refuse can be
utilized as a fuel source in two manners: (1) the direct burning
of refuse results in the release of heat for the production of steam.
Usually the refuse is used to supplement another fuel, and the
refuse provides only a part of the total heat output. (2) the
refuse is converted into a fuel and this fuel is then burned to
produce heat.[2]

In Europe, the use of solid waste as an energy source
is fairly common. This is due primarily to the scarcity of land
necessary for landfills. Also, the availability of refuse for

electric power generation is made possible because in most
countries the same governmental agency has responsibility for both
refuse disposal and electricity generation.

One of the most publicized projects in which refuse
has been used as a fuel is the combined venture of the city of
St. Louis, Union Electric Company and the U.S. Environmental
Improvement Agency. The project utilizes the refuse of St. Louis
at the Union Meramec Power Plant of the Union Electric Company.
The waste materials are domestic in general; they are received at
the processing station and passed through a milling operation to
obtain a uniform size of approximately 1 inch.

An air classifier at the processing plant separates the
less dense, uncombustable materials from the heavier particles which
will not burn, and an electro magnetic separator then removes the
metallic materials. The noncombustible material amounts to about
20 percent of the refuse weight.

Further applications of solid waste utilization in the
United States are now being explored. The Horner and Shifrin
consulting engineering firm has been evaluating the results of the
St. Louis power pilot plant which have been encouraging as to the
possible application of the process to the facilities of the Consoli-
dated Edison Company in New York City. Sometime in the near future
Consolidated Edison may be burning the refuse generated by New York
City.

The U.S. Bureau of Mines' College Park Metallurgy
Research Center in Maryland has developed and is operating a

sophisticated five ton per hour pilot plant which separates refuse

into its basic components. The process distinguishes between chopped

paper, chopped plastics, shredded light iron, heavy iron, heavy

nonferrous metal, chopped aluminum, organic wastes, and glass.

This is an important step forward in solid waste technology for many

of the problems with solid waste disposal incineration and burning

as a fuel are associated with the inadequacies of separation methods

to date.

The Solid Waste Problem

From a puritanical viewpoint a solid waste is everything

that is solid and wasted, rather than recovered in one form or another

by such processes as recycling, conversion, transformation or reuse.

Some common solid waste terms will first be reviewed.

Waste   The word waste refers to useless, unused, unwanted, or

discarded materials. Waste includes solids, liquids and gases.

The gases are principally industrial fumes and smoke; the liquids

consist mainly of sewage and the fluid part of industrial wastes;

the solids are classified as refuse.

Refuse   The term refuse refers to solid wastes. Its component

materials can be classified in several different ways. The point

of origin is important in solving some problems, so that classifying

as plastic institutional, commercial, industrial, street, demolition

or construction is useful. For other problems, the point of origin

is not as important as the nature of the material and classification may be made on the basis of organic or inorganic character, combustibility or noncombustibility, putrescibility or nonputrescibility. One of the most useful classifications is based on the kinds of materials: garbage, rubbish, ashes, street refuse, dead animals, abandoned vehicles, industrial wastes, demolition wastes, construction wastes, sewage solids, and hazardous and special wastes. (See Table 1 for groups of refuse materials and sources.)

Garbage Garbage is the animal and vegetable waste resulting from handling, preparation, cooking, and serving of foods. It is composed largely of putrescible organic matter and its natural moisture; it includes a minimum of free liquids. Garbage originates primarily in home kitchens, stores, restaurants, and other places where food is stored, prepared or served. It should be pointed out that garbage decomposes rapidly, particularly in warm weather, and is a breeding source for flies and vermin, a source of food for rats and produces disagreeable odors. Garbage may have some commercial value as a base for commercial animal feed and as animal food.

Rubbish Rubbish consists of a variety of both combustible and non-combustible solid wastes from homes, stores and institutions. It can be broken down more specifically as combustible and noncombustible rubbish. Combustible rubbish is burnable material and is not highly putrescible which allows it to be stored for long periods. Non-combustible rubbish is material that is unburnable at ordinary incinerator temperatures (1300 to 2000°F). For the most part it consists of inorganic materials such as tin cans, mineral matter, glass, metal furniture, etc.

Ashes   Ashes are residue from wool, coal, coke and other combustibles
burned in homes, stores, institutions and small industrial establishments
for heating, cooking and disposing of combustible waste materials.
Ashes are, in the main, composed of a mixture of fine powdered
material, cinders, clinkers and small portions of unburned or
partially burned material.

Street Refuse   Street refuse is material picked up by manual and
mechanical sweeping of streets and sidewalks and litter from public
receptacles.

Industrial Refuse   Industrial refuse consists of solid waste
materials from factories, processing plants and other manufacturing
enterprises. While not collected by municipal sanitation agencies,
usually on site disposal methods or private collection agencies
employed, they are covered by municipal laws.

Demolition Refuse   Demolition refuse is the waste from razed
buildings or other structures. In most cities, with urban renewal
undertakings, this type of refuse is rising sharply. This type of
refuse is usually collected by private contractors.

Construction Refuse   Construction refuse is the waste material
generated from construction, remodeling, and repair of houses,
commercial buildings and other structures. It is collected, in general,
by private contractors.

                Everyone - producers and consumers - create solid waste.
The larger and more affluent the population, the greater the volume

of solid waste. The problem of what to do with these large masses
is compounded by archaic municipal collection and disposal practices.
However, this problem has existed since Adam discarded the first
apple core. Prehistoric man simply piled his discards in heaps of
shards, bones, and shells and moved on to another habitat when
the garbage masses overwhelmed his cave or small community. The
generation of solid wastes was vastly accelerated with the arrival
of the industrial revolution in the 18th Century. The production
of goods was directly proportional to population and consumption
increases, effecting the enormous quantities of solid wastes that
today are becoming unbearable.

Of the estimated 4 billion tons produced annually
in the United States, animal wastes account for almost two billion
tons; agricultural wastes for nearly 650 million tons; domestic,
commercial and other municipal wastes for about 300 million tons,
and industrial wastes for almost 130 million tons.[3]

Character of Solid Waste

In the early part of the 20th Century, most refuse came
from the kitchen and consisted mainly of food scraps and coal ashes.
Only about fifteen percent of it is now composed of what most people
think of as "real garbage", the sort which is likely to decompose
and putrify. Since World War II, and at an accelerated rate during
the 1950's, convenience packaging has become the hallmark of rapid,
sanitary food preparation. While packaging has reduced the quantity
of garbage and ashes in the former day trash can, such materials as
paper, metals, glass and plastics have become the new components.

Municipal wastes, as we discussed previously, are largely paper and paper products. Table 1 shows the breakdown of municipal wastes into its components and their relative percent by weight. The problem of solid wastes is what to do with them.

TABLE 1 - COMPOSITION AND ANALYSIS OF COMPOSITE MUNICIPAL SOLID WASTE[4]

Components	Percent By Weight
Corrugated paper boxes	23.38
Newspaper	9.40
Magazine paper	6.80
Brown paper	5.57
Mail	2.75
Paper food cartons	2.06
Tissue paper	1.98
Wax cartons	0.76
Plastic coated paper	0.76
Vegetable food wastes	2.29
Citric rinds and seeds	1.53
Meat scraps, cooked	2.29
Fried fats	2.29
Wood	2.29
Ripe tree leaves	2.29
Flower garden plants	1.53
Lawn grass, green	1.53
Evergreens	1.53
Plastics	0.76
Rags	0.76
Leather goods	0.38
Rubber composition	0.38
Paint and oils	0.76
Vacuum cleaner catch	0.76
Dirt	1.53
Metals	6.85
Glass, ceramics, ash	7.73
Adjusted moisture	9.05
Total	100.00

Solid waste disposal is becoming an increasingly large
and expensive program in U. S. communities. The United States, by far
the world's largest producing and consuming nation, faces waste
management problems of enormous proportions. Projections are for
the solid waste generated in the metropolitan areas to more than
triple in the year 2000.

The United States population is becoming increasingly
urban, as sixty-five percent of all Americans now live in cities.
Generally, cities opt for the least expensive means of solid waste
disposal - open dumping and open burning. These methods pollute the
air and water, devour valuable land, create fire and health hazards,
and waste natural resources. The estimated water pollution potential
of garbage generated in one year in New Jersey alone is equivalent
to 120 billion gallons of domestic raw sewage.

Americans make up less than ten percent of the world's
population, yet generate and consume 40 percent of its output.
Unlike trees which are carefully cut, replanted, and scheduled to
insure a constant supply from generation to generation, the supplies
of such resources as petroleum and certain metals are being drained
from the earth until their reserve will be depleted. However, the
problem of solid waste disposal would still be with us even if the
reservoir of world resources were limitless. A common fallacy is
that the "energy crisis" has effected a solid waste crisis, but
in reality, the energy crisis has just made us aware of alternate
energy sources a few years sooner than we expected.

Availability of Solid Wastes

      Resource recovery is a term describing the extraction and
utilization of materials and values from the solid waste stream.
Material recovered such as metals and minerals can be recycled in the
manufacture of new products. Recovery of values could include energy
recovery by utilizing components of waste as fuel; production of
compost using processed solid waste as the medium; reclaiming
fibers, and reclamation of land through sanitary landfill.

      Table 2 shows the value of some products recovered from
one ton of raw refuse.

TABLE 2 - VALUE OF PRODUCTS FROM 1 TON OF RAW REFUSE[3]

Product	Value	Lbs. Per Ton Of Refuse	Value Per Ton Of Refuse
Iron	$10/ton	154	$0.77
Aluminum	$0.12/lb	8	.96
Copper, lead, zinc	$0.19/lb	6	1.14
Glass cullet, color sorted	$12/ton	128	.77
Combustibles (fuel)	$1.23/ton	1,628	1.00
Dirt	NAp	34	-
Waste fine glass	NAp	42	-

Enormous quantities of organic wastes are produced each
year in the United States. The total amount is in excess of
2 billion tons and at least 880 million tons of this is moisture
and free organic material (dry organic solids), representing a
potential energy source of significant magnitude. The most
abundant waste materials containing organic solids are manure,
urban refuse, and agricultural wastes.

Manure

The total annual quantity of animal wastes has been
estimated to be at least 1.7 billion tons. Upon evaluation
of the heating values of these materials the water content must be
neglected. Thus, the actual amounts of organic solids convertible
into oil or gas for fuel are somewhat smaller than the above
figure. Table 3 summarizes the waste generation by major farm
animals. One problem associated with utilizing and estimating the
wastes generated by farm animals is the fact that waste production
can vary with respect to animal and breed type, conditions of
confinement, type of feed, etc.

Agricultural Crop Wastes

The solid wastes generated by major agricultural crops
in 1968 were estimated to be 2,280 million tons. Cornstalks,
pea vines, sugarcane stalks, leaves, stubble, prunings and similar
entities from other plants constitute most of this waste., Much of

TABLE 3 - SOLID WASTE GENERATION BY MAJOR FARM ANIMALS[4]

Animal	Number on Farms (thousands)	Waste Load (Manure)	
		(tons/unit/yr)	(thousand tons/yr)
Cattle	108,862	10	1,088,620
Hogs	47,414	8	379,312
Sheep	21,456	3	64,368
Horses, mules	no estimate since 1960	- - - - -	- - - - - -
Poultry			
Broilers	2,568,338	.0045	11,557
Turkeys	115,507	.025	2,888
Layers	339,921	.047	15,976
Ducks, etc.	no estimate	- - - - -	= - - -- -
TOTAL			1,562,721
			7,814 tons/cap/yr
			15,627.2 lbs/cap/yr
			42.8 lbs/cap/day

this material is now burned to prevent the spread of plant
diseases, or is scattered on farmland to prevent wind and water
erosion.

Commercial

Crushed canestalks (bagasse) from the sugar industry
are burned at the sugar mills to generate steam, and other

agricultural wastes from processing are burned. Wastes such as
stalks and cobs from corn and milling residue from wheat, rice and
other grains are currently incinerated with the heat generated
lost to the atmosphere. It is estimated that 70 percent of these
crop wastes are organic solids which could with the proper technology
contribute to our energy needs. Figure 5 shows the typical agricultural
waste sources of which Table 4 gives the components.

Urban Refuse

          The domestic, municipal and commercial components of urban
waste amount to 3.5, 1.2 and 2.3 pounds per capita per day, respectively.
Quantities of this waste have been estimated to be 260 million tons,
with at least half being dry organic material. At present, approximately 90 percent of this refuse is disposed of in landfills, while

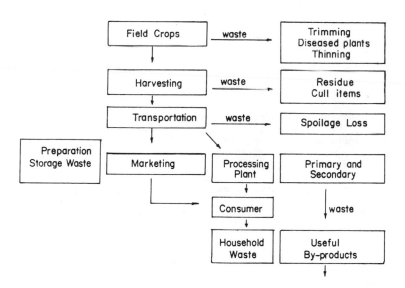

FIG. 5. General agricultural waste sources.[5]

TABLE 4   - COMPOSITION OF SOLID WASTES FROM AGRICULTURAL SOURCES [4]

Agricultural Source	Waste	Composition
Farms, ranches, greenhouses, live-stock feeders, and growers	Crop residue	Cornstalks, tree prunings, pea vines, surgarcane stalks (bagasse), green drop, cull fruit, cull vegetables, rice, barley, wheat and oats stubble, rice hulls, fertilizer residue
	Forest slash	Trees, stumps, limbs, debris
	Animal manure (Paunch manure)	Ligneous and fibrous organic matter, nitrogen, phosphorus, potassium, volatile acids, proteins, fats, carbohydrates
	Poultry manure	Same as animal manure
	Animal carcasses, flesh, blood, fat particles, hair, bones, oil, grease	Ammonia, urea, amines, nitrates, inorganic salts, various organic and nitrogen-containing compounds
	Pesticides, insecticides, herbicides, fungicides, vermicide and microbicide residues and containers	Chlorinated hydrocarbons, organo-phosphorus compounds, other organic and inorganic substances, e.g., strychnine and lead arsenate

most of the remainder is burned in municipal incinerators. Landfill and incineration operations are criticized for their contributions to air and water pollution and because available space
for landfills is scarce. Conversion of urban refuse into some
sort of energy source would appreciably solve solid waste disposal
problems, and the energy value represented by this waste is
constantly renewable. Table 5 shows the composition of solid
wastes from urban sources.

Logging and Wood Manufacturing Residues

Approximately one-third of the volume of wood harvested
in the U.S. is unused. This amounts to over 4.4 billion cubic
feet of wood annually of which 2 billion cubic feet are logging
residues. Some of these wastes are concentrated at sawmills and
veneer mills, and they could be a source of raw material for
conversion to oil or gas, which have a much greater heating value
than wood.

Sewage Solids

Municipal sewage contains organic solids that are
filtered and treated for removal. Sewage, however, is principally
water. Municipal sewage is 99.8 percent water with the remaining
materials in suspension or solution. The treatment of sewage for
the removal of this 0.2 percent generates almost 12 million tons
or organic solids annually - another potentially enormous source
of energy.

Industrial Wastes

Industrial wastes include those wastes from product
processing, packaging, and shipping, including office clerical
refuse. Since industries vary so much in the type and quantity
of materials handled and processed, the composition of wastes from
one industry to another may be completely different. Sludges,
waste plastic, rags, paper and cardboard, scrap metals, slag, rubber,
ceramics, etc., may be generated from a single industrial operation.
The waste generated by industry has been estimated to be 110 million

TABLE 5 - COMPOSITION OF SOLID WASTES FROM URBAN SOURCES[4]

Urban Sources	Waste	Composition
Domestic, household	Garbage	Wastes from preparation, cooking and serving of food; market wastes from handling, storage, and sale of food
	Rubbish, trash	Paper, cartons, boxes, barrels, wood, excelsior, tree branches, yard trimmings, metals, tin cans, dirt, glass, crockery, minerals
	Ashes	Residue from fuel and combustion of solid wastes
	Bulky wastes	Wood furniture, bedding, dunnage, metal furniture, refrigerators, ranges, rubber tires
Commercial, institutional, hospital, hotel, restaurant, stores, offices, markets	Garbage	Same as domestic
	Rubbish, trash	Same as domestic
	Ashes	Same as domestic

Source	Type	Composition
Municipal, streets, sidewalks, alleys, vacant lots, inciner- ators, power plants, sewage treatment plants, lagoons, septic tanks	Demolition wastes, urban renewal, expressways	Lumber, pipes, brick masonry, asphaltic material and other construction materials from razed buildings and structures
	Construction wastes, remodel- ing	Scrap lumber, pipe, concrete, other construction materials
	Special wastes	Hazardous solids and semiliquids, explosives, pathologic wastes, radio- active wastes
	Street refuse	Sweepings, dirt, leaves, catch basin dirt, contents of litter receptacles, etc.
	Dead animals	Cats, dogs, horses, cows, marine animals, etc.
	Abandoned vehicles	Unwanted cars and trucks left on public property
	Fly ash, incinera- tor residue, boiler slag	Boiler house cinders, metal scraps, shavings, minerals, organic materials, charcoal, plastic residues
	Sewage treatment residue	Solids from coarse screening and grit chambers, and sludge from settling tanks

tons annually with 40 percent of this total being organic wastes.
Since the quantity and composition of this waste varies from one
operation to another, heating values of these wastes will vary from
location to location making available industrial waste energy only
to those areas whose waste has a high heat output. Table 6 shows
typical sources and types of industrial wastes.

Miscellaneous Organic Wastes

        Other organic wastes include excretions and carcasses
from cats, dogs, horses and marine animals, organic wastes from
from federal installations, etc., which have been estimated at 50
million tons per year.

Energy Potential from Organic Wastes, 1971 & 1980

        The total quantity of organic wastes generated in the
United States in 1971 is summarized in Table 7.

        The quantities given for 1980 are based on a population
of 236 million. Some increases, which are not caused by population
increases, are expected. These include an annual increase in the
per capita consumption of beef from 112 lbs. in 1970 to 130 lbs. in
1980. The amount of urban refuse per capita is also expected to
increase to at least 8 lbs. per day.

        The significant amount of oil that could be produced
from organic wastes is also given in Table 7. For 1971 the oil
demand in the U.S. was estimated to be 5.8 billion barrels, of
which 1.4 billion barrels were imported. The potential from
organic wastes (1.098 billion barrels/year) is approximately 15

TABLE 6 - SOURCES AND TYPES OF INDUSTRIAL WASTES[4]

Group Classification	Waste Generating Processes	Expected Specific Wastes
Plumbing, heating, air conditioning Special Trade Contractors	Manufacturing and installation in homes, buildings, and factories	Scrap metal from piping and ductwork; rubber, paper, and insulating materials, misc. construction and demolition debris
Ordnance and accessories	Manufacturing and assembling	Metals, plastic, rubber, paper, wood, cloth, and chemical residues
Food and kindred products	Processing, packaging, and shipping	Meats, fats, oils, bones, offal vegetables, fruits, nuts and shells, and cereals
Textile mill products	Weaving, processing, dyeing, and shipping	Cloth and fiber residues
Lumber and wood products	Sawmills, mill work plants, wooden container, misc. wood products, manufacturing	Scrap wood, shavings, sawdust; in some instances metals, plastics, fibers, glues, sealers, paints, and solvents
Apparel and other finished products	Cutting, sewing, sizing, and pressing	Cloth and fibers, metals, plastics, and rubber
Furniture, wood	Manufacture of household and office furniture, partitions, office and store fixtures, and mattresses	Those listed under Lumber and Wood products and in addition cloth and padding residues
Furniture, metal	Manufacture of household and office furniture, lockers, bedsprings, and frames	Metals, plastics, resins, glass, wood, rubber, adhesives, cloth, and paper
Paper and allied products	Paper manufacture, conversion of paper and paperboard, manufacture of paperboard boxes and containers	Paper and fiber residues, chemicals, paper coatings and fillers, inks, glues, and fasteners
Printing and publishing	Newspaper publishing, printing, lithography, engraving, and book-binding	Paper, newsprint, cardboard, metals, chemicals, cloth, inks, and glues

TABLE 6 - SOURCES AND TYPES OF INDUSTRIAL WASTES[4]   (cont'd)

Industry	Source/Process	Types of wastes
Chemicals and related products	Manufacture and preparation of inorganic chemicals (ranges from drugs and soups to paints and varnishes, and explosives)	Organic and inorganic chemicals, metals, plastics, rubber, glass, oils, paints, solvents and pigments
Petroleum refining and related industries	Manufacture of paving and roofing materials	Asphalt and tars, felts, asbestos, paper, cloth, and fiber
Rubber and miscellaneous plastic products	Manufacture of fabricated rubber and plastic products	Scrap rubber and plastics, lampblack, curing compounds, and dyes
Leather and leather products	Leather tanning and finishing; manufacture of leather belting and packing	Scrap leather, thread, dyes, oils, processing and curing compounds
Stone, clay, and glass products	Manufacture of flat glass, fabrication or forming of glass; manufacture of concrete, gypsum, and plaster products; forming and processing of stone and stone products, abrasives, asbestos, and misc. nonmineral products	Glass, cement, clay, ceramics, gypsum, asbestos, stone, paper, and abrasives
Primary metal industries	Melting, casting, forging, drawing, rolling, forming, and extruding operations	Ferrous and nonferrous metals scrap, slag, sand, cores, patterns, bonding agents
Fabricated metal products	Manufacture of metal cans, hand tools, general hardware, non-electric heating apparatus, plumbing fixtures, fabricated structural products, wire, farm machinery and equipment, coating and engraving of metal	Metals, ceramics, sand, slag, scale, coatings, solvents, lubricants, pickling liquors

Machinery (except electrical)	Manufacture of equipment for construction, mining, elevators, moving stairways, conveyors, industrial trucks, trailers, stackers, machine tools, etc.	Slag, sand, cores, metal scrap, wood, plastics, resins, rubber, cloth, paints, solvents, petroleum product
Electrical	Manufacture of electric equipment, appliances, and communication apparatus, machining, drawing, forming, welding, stamping, winding, painting, plating, baking, and firing operations	Metal scrap, carbon, glass, exotic metals, rubber, plastics, resins, fibers, cloth residues
Transportation equipment	Manufacture of motor vehicles, truck and bus bodies, motor vehicle parts and accessories, aircraft and parts, ship and boat building and repairing motorcycles and bicycles and parts, etc.	Metal scrap, glass, fiber, wood, rubber, plastics, cloth, paints, solvents, petroleum products
Professional, scientific controlling instruments	Manufacture of engineering, laboratory, and research instruments and associated equipment	Metals, plastics, resins, glass, wood, rubber, fibers, and abrasives
Miscellaneous manufacturing	Manufacture of jewelry, silverware, plated ware, toys, amusement, sporting and athletic goods, costume novelties, buttons, brooms, brushes, signs, and advertising displays	Metals, glass, plastics, resins, leather, rubber, composition, bone, cloth, straw, adhesives, paints, solvents

TABLE 7 - ESTIMATES OF ORGANIC WASTES GENERATED, 1971 and 1980[6]

SOURCE			1971	1980
Manure	Million Tons/Year		200	266
Urban Refuse	"	"	129	222
Logging and wood manu- facturing residues	"	"	55	59
Agricultural crops and food wastes	"	"	390	390
Industrial wastes	"	"	40	50
Municipal Sewage Solids	"	"	12	14
Miscellaneous Organic Wastes	"	"	50	60
TOTAL			880	1,061
Net Oil Potential	Million Barrels		1,098	1,330
Net Gas for Fuel Potential	Trillion Cubic Feet		8.8	10.6

percent of the 1971 oil demand and over 77 percent of the oil imported in 1971. A likely use of this oil would be as fuel oil for the generating of electric power, which would supplement the usual supplies from petroleum. The domestic residual fuel oil demand in 1971 was approximately 860 million barrels.

We also note the quantity of gas that could have been produced from the quantities of wastes shown. In 1971 the 8.8 trillion cubic feet would have amounted to over 38 percent of the 22.8 trillion cubic feet used in the U.S.

Incineration Potential of Municipal and Industrial Waste

Both municipal and industrial wastes possess relatively high heating value, which allows them to be incinerated with very little, if any, additional fuel needed. Figure 6 shows the composition limits for self-burning refuse. Within these limits, no additional fuel, such as oil or gas, is needed.

Incineration is a solution that can reduce the amount of refuse for disposal to about one-tenth its original volume. The heat given off by the incineration of refuse can be used to make steam which can be used for power generation or district heating purposes.

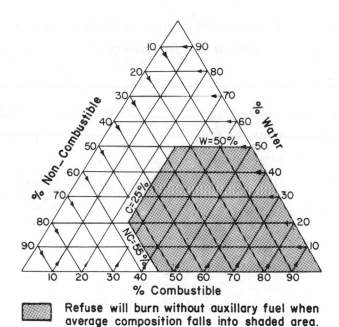

Refuse will burn without auxiliary fuel when average composition falls into shaded area.

FIG. 6. Composition limits for self-burning refuse.[7]

The per capita production of municipal waste was approximately one ton per person. Table 8 shows the moisture contents and
heating values of combustible municipal refuse components. Composite
municipal waste generally has a heating value between 4,500 and
6,500 BTU/lb. Heating values, in general, can be raised by some
type of drying process.

The present economic and energy shortage situation has
effected a trend of urban refuse disposal and volume reduction
through power generation. An example of this is Riverside, California
which will purchase electricity generated from solid waste combustion
at a one million dollar pilot plant engineered by Pyrolysis Systems,

TABLE 8 - TYPICAL MOISTURE CONTENT AND HEATING VALUES OF
MUNICIPAL REFUSE COMPONENTS[5]

	Moisture %	BTU/lb.
Paper, Cardboard, Cartons, Bags	3	7,660
Wood crates, Boxes, Scrap	7	7,825
Brush, Branches	17	7,140
Leaves	30	4,900
Grass	50	3,820
Garbage	75	1,820
Greenstuff	50	3,470
Greens	50	4,070
Rags, Cotton, Linen	10	6,440

Inc., Riverside. The plant will convert 50 tons of urban wastes
per day to a fuel gas to generate steam and electricity. Table 9
shows the gas product's composition. This gas product is the energy
source of a heat exchanger which effects the pyrolytic decomposition.[8]

Numerous industrial wastes have appreciable heating values.
Table 10 indicates several "valuable" industrial wastes and their
respective heating values. Composite industrial solid waste generally
has a heating value between 6,600 and 7,300 BTU/lb.

Industrial wastes are different from municipal wastes
in that they can be any imaginable material, and a solid industrial
waste may be that in name only. Thus the characteristics of a waste

TABLE 9 - GAS PRODUCT COMPOSITION FROM MUNICIPAL
SOLID WASTE PLANT[8]

Component	Percent
Hydrogen	19
Methane	17
CO	25
CO_2	15
Nitrogen	15
Propane and Butane Ethylene	9

TABLE 10 - LISTING OF COMMON WASTES WITH FUEL VALUE (AVERAGE OR
AVERAGE RANGE)[9]

	BTU/LB, AS FIRED
GAS	
Coke-oven gas	19,700
Blast-furnace gas	1,139
CO gas	575
Refinery gas	21,800
LIQUID	
Industrial sludge	3700-4200
Black liquor	4,400
Sulfite liquor	4,200
Dirty solvents	10,000-16,000
Spent lubricants	10,000-14,000
Paints and resins	6,000-10,000
Oil waste, fuel oil residue	18,000
SOLID	
Bagasse	3,600-6,500
Bark	4,500-5,200
General wood wastes	4,500-6,500
Sawdust and shavings	4,500-7,500
Coffee grounds	4,900-6,500
Nut hulls	7,700
Rice hulls	5,225-6,500
Corn cobs	8,000-8,300
Boot, shoe trim and scrap	8,500
Sponge waffle and scrap	8,500
Butyl soles scrap	11,500
Cement wet scrap	11,500
Rubber	~12,420
Tire cord scrap	~12,400
Tires, bus and auto	~18,000
Gum scrap	~19,700
Latex waste, coagulum waste	~12,000
Leather scrap	10,000
Waxed paper	~12,000
Cork scrap	12,400
PLASTIC AND SYNTHETIC REFUSE	
Cellophane plastic	12,000
Polyethylene	19,840
Polyvinyl chloride	17,500
Vinyl scrap	17,500
Aldehyde sludge	18,150
Solvent naptha	18,500
Carbon disulfite	8,000
Benzine	10,000

must be carefully defined before an application as an energy source
can be chosen. The following should be determined.

1. Moisture content

2. Volatile matter content

3. Fixed carbon content

4. Ash content

5. Heating value

6. Corrosiveness

7. Toxicity

8. Odor

9. Explosiveness

10. Flash point

11. Density

12. Ash-fusion temperature

Processing solid wastes, as we shall see, can become very complex as
we search for the most economical, yet efficient system. A solid
waste disposal, volume reduction, and combustion system must consider
many factors including:[5]

1. Segregation

2. Transportation

3. Solid and liquid preparation

4. Storage and equalization

5. Handling and feeding

6. Combustion process requirements

7.   Corrosion or other facility damage

8.   Residue handling and disposal

9.   Environmental impact

CHAPTER 2

ECONOMIC CONSIDERATIONS

The economic feasibility of energy recovery from solid waste is a fairly complex problem which requires a broad and detailed study. The major economic elements which must be studied include:

. The comparable unit costs for alternate means of solid waste disposal

. The total unit cost of processing and transporting the prepared solid waste to the point of use

. The unit costs of firing the solid waste

. The net value of the milled solid waste as a replacement for the normal fossil fuel

. The value of magnetic metals and other materials which potentially can be recovered from the non-burnable fraction

Since these factors will vary with each application, the evaluation of a potential solid waste energy recovery unit must be considered as a unique entity. Aside from rising costs the evaluation should include a complete analysis of the reliability of

the solid waste quantity and a realistic assessment of quantity of

the organic fraction which might be utilized as a fuel.[10]

Presently the most common method of disposal entails

operation of a sanitary landfill relatively close to the solid waste

source origin, resulting in a cost for disposal of about three to five

dollars per ton. These figures are the national average excluding

transportation costs. However, most urban sanitary landfills are

saturated, effecting the need to haul garbage further and further

from its source.[11]

Alternate disposal methods are becoming not only economically

feasible, but environmentally mandatory. Landfill sites are minimal

because the amount of available land is diminishing, governmental

regulations require specific standards for air, water, and pollution,

and no one wants a landfill, sanitary or otherwise, in their backyard

anymore. There would also be some energy savings in the 287 million

gallons of gasoline and 163 million gallons of diesel fuel consumed by

the 100,000 garbage trucks of our nation each year. A centrally

located disposal complex could drastically cut fuel consumption.

Support for Solid Waste Conversion

Although solid wastes have high heating values, they

cannot be utilized as economically and easily as uniform fuels with

concentrated energy contents such as natural gas, oil, or coal.

However, the present cost for solid waste disposal also has an

economic value. By burning solid wastes for fuel we create an

energy source and disposal method with one act. Thus, some of the
costs of collection, transportation, and processing can be considered
as the fuel price. Ordinarily this whole cost would support the
price of solid waste disposal. The fractional cost of solid waste
energy recovery that is above the cost of conventional fuels can be
considered the disposal cost which would be far below present methods.
The economic feasibility of an energy conservation system utilizing
solid wastes will vary with location and as inflation continues,
with time. It is most significant to point out that as fuel costs
increase the larger portion of the dual support will come from the
fuel value of the solid waste and the cost attributed to disposal
can then be decreased. Presently, the total operating costs and
capital charges for deriving the energy from solid waste per ton
are in the range of $15. to $20. Relating the heat extracted from
solid waste to fuel oil of equivalent heat content at a cost of
$6. a barrel, the value of energy derived from a ton of refuse is about
$9. assuming a high utilization efficiency. However, if a more realistic
lower efficiency is considered for the conversion of this energy into
useful power, a lesser value of $6. per ton of refuse is obtained. Using
this conservative estimate, in order to obtain profits from sales of
this energy, the cost required for disposal-energy recovery is between
$9. and $14. per ton.[12] This cost is relatively close to the present
cost for distant landfill sites. Figure 7 illustrates the impact of
changing fuel oil prices on the necessary disposal charge, assuming
fixed processing costs.

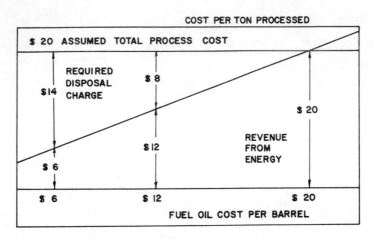

FIG. 7.   Increasing fuel value lower disposal charges.

Two Basic Systems

A comparative economic study of the many different types
of systems designed to recover energy from solid waste results in
two basic categories:

1)   Systems of low initial investment and high operating
costs.

2)   Systems of initial investment and generally lower
operating costs.

Typical systems with high operating costs include plants
that shred incoming solid waste and, at a minimum, separate the many
components at the onset by means of a system of screens and/or air
classifiers to produce a fuel. High capital investment systems

derive the maximum available energy through combustion of the
solid waste in specially designed steam generating units essentially
as the waste is received.

Because the fuel producing, shredding, and separating
plants do not themselves produce energy or use boilers they are
not as capital intensive as those systems that use specialized
boilers in order to generate steam. However, the maintenance
and operating costs for shredding, separating and classifying equipment
is a significant, continuous expense, and transportation of the
fuel usually becomes necessary. As compared to the total combustion
systems that consume solid waste as received without prior processing,
these systems are operating cost intensive.

When evaluating the comparative economics of the two basic
systems for energy recovery in relatively similar operating circumstances,
the total average annual cost over the life of the plant is
important. Annualized capital as well as plant operating costs
must be considered. Capital costs include amortization of the
initial investment with respect to expected life and interest and
return rates. Operating costs include labor, administration,
maintenance and repairs, utilities, taxes, and a multiplier to be
used as an inflation factor to adjust predicted operating costs
for the life of the project.[12]

In cost comparisons it is important to recognize that a
unit of heat, or BTU, as steam is more valuable to a consumer

than the potential BTU in a fuel. This is because the potential
BTU in a fuel must be extracted in a furnace-boiler system and such
a process requires a capital investment and associated expenditures
for boiler plant operations. This factor should be considered even
if solid waste is used as a complete or partial substitute fuel in
an existing boiler. A comparison has been made between a capital
intensive steam or hot water generating plant using 100% unshredded
refuse and a plant producing shredded refuse-derived fuel. Without
adjustment, the shredded fuel plant may appear to be somewhat less
costly per process input ton. However, an evaluation of the cost
between steam and fuel is not strictly valid without some adjustment.
Converting the shredded fuel to steam requires the use of a boiler even
if it is existing, and some allowance must be made for the operative
and maintenance costs prorated on the amount of solid waste used.

     These should include costs for the following:

1) boiler feed   (conveyor or blower)

2) draft induction

3) pumping feed water

4) ash extraction

5) additional manpower effected by the burning of two
widely diverse fuels

    6)   additional maintenance resulting from:

        a)   corrosion or slagging

        b)   air pollution controls and related problems

A conservative estimate of the total operating cost of taking the shredded refuse and converting it to steam in an existing boiler would be a minimum of $1.50 per ton of original solid waste.[12] Obviously under today's economic conditions, where we no longer compare 1960 dollars with 1970 dollars, but compare 1975 dollars to 1976 dollars, operating costs and investment estimates have become critical. Therefore, each municipality must carefully evaluate the options, the benefits, and the drawbacks of each system and then decide upon the process best suited to its individual needs.

Table 11 compares the economics of 11 different systems for solid waste disposal and the cost differential resulting from the units being municipally or privately owned. In Figure 8 these 11 systems are examined as to the variation of costs with respect to the size of the operation. All show a decrease in unit costs as plant size increases.

TABLE 11 - ECONOMIC EVALUATION OF SYSTEMS[13]

Systems	*Municipally Owned $/Ton	Privately Owned $/Ton
Sanitary Landfill (close in)	2.57	3.04
Fuel Recovery (prepared as a suitable boiler fuel)	2.70	4.30
Materials Recovery	4.77	7.21
Pyrolysis	5.42	8.02
Sanitary Landfill (remote)	5.94	6.47
Composting (mechanical)	6.28	9.91
Incineration Steam and Residue Recovery	6.57	9.27
Incineration and Steam Recovery	7.05	9.50
Conventional Incineration + Residue Recovery	7.18	9.43
Conventional Incineration Only	7.68	9.64
Incineration and Energy Recovery (electrical generation)	8.97	12.72

*Since most waste reclamation systems require high capital investments per unit of plant capacity, municipally owned plants offer significant cost advantages over privately owned plants. Tax-free municipal bonds can be sold at about a 7 percent rate while industrial bonds yield 1 percent. Municipalities also have no property taxes or income taxes to contend with.

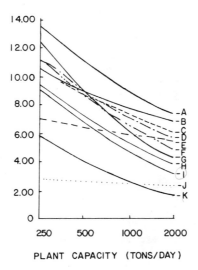

PLANT CAPACITY (TONS/DAY)

FIG. 8. Net operating costs associated with municipally owned resource recovery processes at various plant capacities (20 year economic life; 300 days per year operation).[14] A, Incineration and electric generation; B, incineration only; C, incineration and residue recovery; D, incineration and steam recovery; E, remote sanitary landfill; F, incineration, steam and residue recovery; G, composting mechanical; H, pyrolysis; I, materials recovery; J, sanitary landfill; K, fuel recovery = supplemental boiler fuel ($0.25/million BTU).

Table 12 is a summary of comparative capital costs between the Saugus, Mass. and the Horner & Shifrin systems. Table 13 then summarizes the operating costs of two systems. These tables show that the combined operating costs for these two systems are comparable. These costs are $13.98 per ton of solid waste for the Saugus, Mass. facility and $13.10 per ton of solid waste for the Arthur Kill Horner & Shifrin system.

TABLE 12 - SUMMARY OF CAPITAL COSTS[15]

---

100 PERCENT REFUSE FIRED BOILER (Based on Installation at Saugus, Mass.)

		Annual Cost (Note 1)	Per Ton (Note 2)
Basic Plant	$27,000,000		
Energy Trans. Syst.	400,000		
Boiler Mod.	400,000		
	$27,800,000	$3,265,388	$7.46

SHREDDED FUEL (Horner & Shifrin estimate, unadjusted)

Shredder Plant	$ 5,281,000
Fuel Trans. Syst.	
and Storage	2,805,000
Boiler Mod.	1,280,000
	$ 9 366,000
Less Building and	
Structural	- 300,000
	$ 9,366,000

AMORTIZATION

Amount	Factor		Annual Cost	Per Ton
$ 300,000	0.1102 (Note 3)		$ 33,060	$0.08
$ 9,066,000	0.1845 (Note 4)		$1,672,677	$3.82
		TOTAL	$1,705,737	$3.90

Note 1. Amortization factor for 10%, 20 years.   0.11746

Note 2. For 438,000 tons/yr

Note 3. Amortizing building and structural cost over 25 years @ 10% cost
of capital

Note 4. Amortizing processing, transport, and injection equipment over 20 years,
for 10 year life, 10% cost of capital and second purchase price
inflated at 4% per year for 10 years.

TABLE 13A - SUMMARY OF OPERATING COSTS

100% REFUSE FIRED BOILER SYSTEM[15]

(Based on Saugus System)

	Annual Cost	Cost Per Input Ton	
ENERGY PRODUCTION			
Administration	$100,000	$0.23	
Labor	715,000	1.63	
Maintenance and Parts	725,000	1.66	
Utilities	407,000	0.93	
	$1,947,000		$4.45
ENERGY DELIVER TO UTILITY (Included in cost of energy production)			
COST AT UTILITY	75,000		0.17
RESIDUE & ASH DISPOSAL	258,000		0.59
TAXES & INSURANCE (@ 2% of Capital)	576,000		1.31
TOTAL	$2,856,000		$6.52

TABLE 13B - SUMMARY OF OPERATING COSTS[15]

SHREDDED FUEL SYSTEM

(Based on Arthur Kill - H & S Report, Adjusted to 438,000 Tons/Yr)

	Annual Cost		Cost Per Input Ton	
PROCESSING				
Administration	$100,000		$0.23	
Labor	760,000		1.74	
Utilities	400,000		0.91	
Maintenance	437,000		1.00	
Parts Replacement	171,000		0.39	
Supplies	60,000		0.14	
		$1,928,000		$4.41
FUEL DELIVERY TO UTILITY				
Administration	5,000		0.01	
Labor	40,000		0.09	
Utilities	278,000		0.63	
Maintenance	110,000		0.25	
Equipment Replacement	19,000		0.04	
Supplies	6,000		0.01	
		458,000		1.93
COST AT UTILITY - FUEL STORAGE & HANDLING				
Operating Labor	250,000		0.57	
Maintenance	70,000		0.16	
Utilities	44,000		0.10	
Allowance for Unknowns	87,000		0.20	
		451,000		1.03
PRORATED BOILER OPERATING COST				
Fuel to Steam		657,000		1.50
RESIDUE & ASH DISPOSAL				
Shredding Residue	219,000		0.50	
Ash	131,000		0.30	
		350,000		0.80
TAXES & INSURANCE (@ 2% of Capital)		187,000		0.43
TOTAL		$4,031,000		$9.20

THE ST. LOUIS SOLID WASTE
DEMONSTRATION PROJECT

The generation of electricity through the burning of solid
wastes has been part of the European scene for a number of years.
The United States, rich in natural resources, has not seen the need
until recently to develop alternate energy sources. Foreign oil
prices are expected to reach $20 per barrel by mid-1978 and with this
Americans can no longer afford to waste valuable resources just because
recovery is uneconomical. Resource recovery costs from solid wastes
are rapidly becoming compatible with other energy sources, not only
because of rising fuel oil prices, but as monies are poured into
research and development technology can produce more efficient and
less costly methods for recovery.

Old inefficient waste-heat boilers installed on old incinerators
have given way to more sophisticated solid waste incinerators built
with steam recovery boilers. Originally these new systems were
neglected as a result of high initial investment and operating costs,
and questionable reliability -- both in performance and markets. How-

ever, utilities that produce electricity already have developed markets, a continuous demand for energy, and boilers which may be modified to accept solid wastes, or the capital to build new boiler facilities.

The St. Louis Project

Since the Spring of 1972, Union Electric Company, in cooperation with the City of St. Louis, has been burning prepared refuse in one of its existing pulverized-fired boilers. The project was partially funded by the Environmental Protection Agency with an initial grant of $3.3 million.

Operation of the project is quite simple as the domestic solid waste collected from the greater metropolitan St. Louis area passes through a large hammermill and is shredded. An air classifier then separates this material into a light combustible fraction and a heavy non-combustible fraction. The combustibles are pneumatically fired into the boiler facilities of the Union Electric Company. Ferrous metals are recovered from the heavy waste fraction, and the remaining heavy residue -- glass, ceramics, and other nonmagnetic materials -- is landfilled.[16]

From the environmental and economical viewpoints this system is desirable:

1)   Conservation of natural resources.

2)   95 percent reduction in solid waste land disposal require-ments.

3)   Air and water pollution are decreased.

4)   Electric power is generated.

5)   Revenues are derived from the sale of ferrous metals and other recovered waste materials.

Union Electric Company examined the question of burning solid wastes for power production. In utilizing a small percentage of properly prepared domestic refuse along with the normal fuel supply, little effect on the boilers is noted. The problems already associated with coal-fired boilers would not be increased with the addition of small amounts of the solid waste material. Furthermore, where domestic solid waste in its original form has a heating content of about 5000 British Thermal Units (BTU) per pound, the light combustible fraction which is being fired has a heating content of 10,000 BTU/lb.

The St. Louis Processing Facilities

Responsibility for processing and transporting the solid wastes is with the City of St. Louis. Only residential solid wastes are accepted for this process since the system was not designed to handle commercial and industrial wastes as well as junked cars and car parts, furniture, large appliances and other bulky wastes. Because of the relatively high heating values of certain industrial wastes it is hoped that these materials may be incorporated into the system in the future.

The capacity of the processing facility is 325 tons per 8-hour shift which is 10 percent of the boiler heat requirements. Maintenance of the processing plant requires a full 8-hour shift as hammertip replacement of the hammermill is a daily function. Longer maintenance times result in the necessity for all such systems to have alternate disposal facilities. Ideally an auxilliary process line would treat the wastes when the primary process line is down. However, to reduce capital costs, the alternate methods opted by St. Louis were extra

storage space for short downtimes and use of the city's existing
incinerator for longer outage periods.

Packer-type collection trucks dump their raw refuse on the
open floor of the receiving building, where it is pushed to a receiving belt conveyor by front end loaders. The solid wastes are transferred
to the hammermill for shredding by the belt conveyor.

The hammermill consists of 30 large metal hammers attached
to a horizontal shaft and 1250 Hp motor. The refuse is ground against
a mill grate with openings 2 1/4 inches by 3 1/4 inches through which
the shredded particles drop. This hammermill is acceptable because
1) it effects a homogeneous effluent from a heterogeneous influent;
2) it meets required and consistent production rates, and 3) it
controls particle size.[16,17]

Over 90 percent by weight of the particles passing through the
milling operation are less than one inch, and nearly all the particles
are less than 1 1/2 inches in size. The optimum particle size with
respect to economics, combustibility, handling and air emissions has
yet to be determined. Depending on the moisture content and composition
the uncompacted bulk density of the particles varies from 4 lb/ft^3
to 12 lb/ft^3. This has caused some problems with equipment designed on
gravimetric and/or volumetric basis.

To improve the milling process and reduce hammer wear, re-
commendations for future applications would include a two-stage miller.
The first run would reduce the waste to 4 to 8 inch particles and the
heavy materials would be removed by the air classifier. The light
fraction would then be reduced to sizes of about 1 or 2 inches.

Figure 9 shows the process flow plan of the solid waste treatment and resource recovery system. The shredded waste is conveyed to the air classifier metering and surge bin which controls the feed to the classifier inlet. Figure 10 is a cross-section of the air classifier.

The classifier separates the light combustible components from the bulk of the milled solid waste. The light fraction is withdrawn from the top of the classifier while the heavier materials are withdrawn from the bottom.

Air is supplied by a 40,000 ft^3/min air foil-induced-draft fan with inlet vane dampers at the bottom (heavy fraction outlet) of the classifier. The inlet vanes and adjustable walls control the flow of air through the separation zone of the classifier. The air flow is adjusted to allow a 75-80 percent separation of light combustibles. The light fraction is withdrawn by air to a cyclone separator, discharged to a conveyor belt, and transported to a storage bin. Clean air is recycled from the cyclone back to the inlet of the classifier fan.

The separation of the solid waste into light and heavy components effects a higher heating value for the fuel, an ease of handling through trucking to furnace injection, a more suitable boiler bottom ash for reuse, and a lesser ash content of the fuel. The heavy fraction is passed under a magnetic belt separator to effect magnetic metals removal. The magnetic metals in the raw form have a density of about 35 lb/ft^3 and are fed into a 100 Hp vertical "nuggetizing" mill which increases their density to about 60-70 lb/ft^3. This material then passes over a magnetic drum to separate the remaining non-magnetic metals. The densified ferrous metals are transported to the Granite City Steel

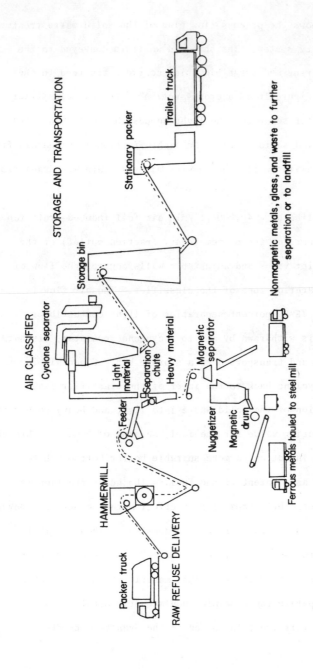

FIG. 9. Solid waste processing facilities.[16]

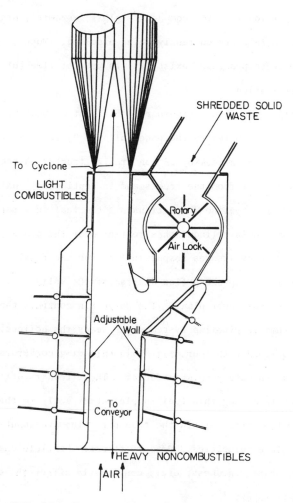

To Cyclone

LIGHT COMBUSTIBLES

SHREDDED SOLID WASTE

Rotary

Air Lock

Adjustable Wall

To Conveyor

HEAVY NONCOMBUSTIBLES

AIR

FIG. 10. Cross section of an air classifier.

Company, Granite City, Illinois, for use as scrap in charging blast
furnaces. The optimum value and use of these magnetics as substitutes
for iron ore in each blast furnace has not yet been ascertained.

    After the removal of the magnetics, the heavy fraction is
landfilled but at a 95 percent volume reduction. New methods to further

separate this portion into its components glass, ceramics, organics,
and non-ferrous metals are currently being examined. Much of the
technology for these processes exists at the present time but
is awaiting application.

The City of St. Louis transports the light combustible fraction
of the solid waste 18 miles to the power plant. Deliveries are made
continuously five days a week by 25 ton trailer trucks. A receiving bin
accepts the waste fuel from the trucks and is continuously unloaded by a
pneumatic transport pipeline which transfers the fuel to a surge bin.
At this point St. Louis relinquinshes control over the process and the
Union Electric Company assumes responsibility for the final operations.

Figure 11 is a process flow diagram of the solid waste fuel as
it passes through the power plant. The surge bin controls the flow of
fuel to the pneumatic pipeline system which feeds the boilers. The
surge bin is equipped with four drag-chain unloading conveyors each
of which feed a separate pneumatic feeder. These feeders utilize a
rotary air lock to convey this fuel supplement directly to the boiler
furnaces. Air flow rates within the transport pipe are about 80 to 90
ft/sec and particle velocities, 50 to 70 ft/sec. Particle characteristics,
size, density, shape, hardness, etc., can greatly affect the conditions
and manner of transport.[18]

Test Boilers

Figures 11 and 12 show the type of boiler being used for this
project. The two modified boilers are relatively small when compared
to the more modern units of the Union Electric Company System, but of
comparable heating design. Each unit is fired tangentially from each

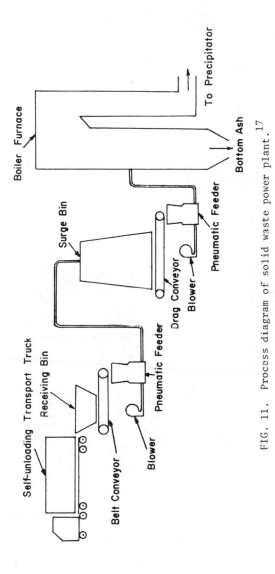

FIG. 11. Process diagram of solid waste power plant.[17]

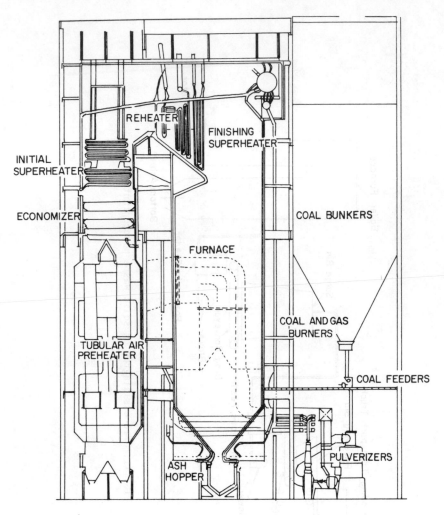

FIG. 12.   The Meramec boiler unit no. 1.

corner of the furnace with four pulverized coal burners. Each burns
nearly 60 tons/hr of bituminous coal at the nominal rate load of 125MW.
The furnace dimensions are about 28 by 38 feet, with an inside height
of about 100 feet.

12.5 tons/hr or 300 tons/24 hr of refuse are equivalent to the
heating value of 10 percent of the coal at full load. The system was
designed around existing collection practices in St. Louis, or to burn
refuse on a 24-hr-a-day, 5-day-a-week cycle because refuse collections
are made on a 5-day-a-week basis. This intermittent firing of the
refuse has had little effect on the boiler operations. Furnace
modifications only entailed the installation of refuse burning ports
at each corner between the two middle tangential coal burners. In
utilizing the same flame pattern as coal or gas the solid waste is burned
in suspension. Those solid waste particles which don't burn or are too
large to be completely combusted during exposure time in the furnace fall
into the bottom ash hopper and are disposed of with the bottom ash.

The processed refuse fires at a constant rate allowing combustion
controls on the boiler to automatically vary the rate of firing of
pulverized coal to maintain the heat requirements of the boiler. Flow
characteristics and heating values of the processed solid waste vary daily.
Complex control systems for refuse feed are needless and cannot be
economically justified. The feed rate may vary by $\pm$ 10 percent or more,
but acceptable compensation is made through the varying of coal input
process. Refuse feed is interrupted occasionally and this is considered
normal operation function.[16,17]

Refuse Analyses

An analysis of the composition of the milled and classified
solid wastes is given in Table 14. Samples were taken for nearly a
year, thus taking into account seasonal variations of the residential

TABLE 14 - REFUSE ANALYSIS PRIOR TO INJECTION INTO
THE BOILERS[16,17]

	Average*	Maximum*	Minimum*
Moisture (%)	30.0	50.0	9.3
Sulfur (%)	0.10	0.40	0.03
Chlorides (%)	0.3	0.95	0.13
Ash (%)	20.0	52.2	12.1
Heating Value, BTU/lb	4589	7425	3027

*Values do not equal 100% because these numbers are relative to their
particular waste sample.

wastes. Depending on the season, the area of collection and weekly
weather conditions, the composition can vary widely and so does the
heating value. Some days require quantities of solid wastes two and
a half times other day's quantities to be burned to effect the same
heat output. This is affected by large variations in moisture and
ash content which can be as high as 50 percent and drop to around
10 percent.

Table 15 shows some typical characteristics of coal as compared
to solid wastes. While the solid wastes' sulfur content is quite low,
the moisture, ash and chlorine contents are relatively high with respect
to typical bituminous coal from southern Illinois. These solid wastes

TABLE 15 - COMPARISON OF THE COMPOSITION OF COAL AND
RESIDENTIAL SOLID WASTES[16]

	Percent of Sample by Weight		Heating Value (lbs per million BTU)	
	Solid Waste	Coal	Solid Waste	Coal
Sulfur	0.1	3-4	0.2	2.6
Ash	20.0	10-11	43.0	9.0
Chlorine	0.3	0.03-0.05	0.6	0.03
Moisture	30.0	6-10	64.0	7.0
BTU per pound	Solid waste: 4,675		Coal: 11,300 to 11,900	

have a heat output of less than 50 percent of similar amount of coal.
However, the analysis shown is for partially processed solid wastes,
i.e. only the magnetic metals were removed following milling. Tests
have shown the light fraction of air classified materials to have
heating values comparable to coal.[10]

The low sulfur content is very interesting from the environmental viewpoint. SO_2 emissions can be greatly reduced upon the
utilization of a 1 percent sulfur fuel. In fact governmental regulations
require the burning of low sulfur fuels in many areas in addition to
pollution control devices. Low sulfur fuels have become expensive and
their supply is limited. The use of solid wastes as a substitute or
supplement to these fuels could greatly reduce power costs to the
consumer. Sulfur emissions from solid wastes are expected to decrease

significantly through further processing. Further treatment of the
solid waste is expected to reduce sulfur content to 0.5 lb per million
BTU of fuel value and thus effect lower SO_2 emissions.

Nitrogen oxides (NO_x) are emitted as a result of the oxidation
of nitrogen in the fuel (fuel NO_x) and in the air needed for combustion
(thermal NO_x). Fuel NO_x's are of no concern because the quantity of
nitrogen in the solid wastes is negligible. Quantities of thermal NO_x
are directly proportional to the temperature of combustion. Solid wastes
burn at a much lower temperature than coal or oil thus inhibiting the
formation of thermal NO_x. Further combustion modifications can completely
eliminate these pollutants.

Particulate emissions are not significantly greater with solid
waste burning as a supplement to coal than with coal burning alone.
Electrostatic precipitators (ESP's) can effectively control the fly ash
of combustion processes. ESP's efficiencies are improved with larger
particle sizes, and solid waste born particulates are generally believed
to be larger than coal born particulates.

However, any increased particulate emissions from solid waste
combustion will probably only increase ESP requirements by 20 percent to
maintain given emission rates.

The Union Electric Company's solid waste process discharges
no wastewater. However, the boiler residue (bottom ash) is deposited
in a settling pond[18] effecting a source of water pollution. Included
in this bottom ash is solid waste ash and an analysis of this residue
is given in Table 16. A further comparison, Table 17, shows the water
analysis of the ash disposal pond, a typical, coal source bottom ash

TABLE 16 - SOLID WASTE BOTTOM ASH ANALYSIS[17]

Component	Average*	Maximum*	Minimum*
P_2O_5	1.38	2.37	0.29
SiO_2	56.4	68.40	38.80
Al_2O_3	7.96	16.30	4.00
TiO_2	0.98	2.37	0.28
Fe_2O_3	4.19	12.16	1.40
CaO	11.13	14.65	5.77
MgO	1.49	2.17	0.79
SO_3	1.60	9.74	0.07
K_2O	2.08	7.62	1.14
Na_2O	7.34	10.92	4.89
SnO_2	0.02	0.085	0.006
CuO	0.24	0.90	0.08
ZmO	0.46	1.91	0.26
PbO	0.25	1.20	0.16

*Percent by weight of dry sample.

pond meeting the Mississippi River and Missouri state standards. The
refuse pond effluent does not significantly differ from other water
effluents and is in general compliance with the regulations.

The control of the environmental impact of solid waste burning is
one of the major stumbling blocks inhibiting the widescale development
of solid waste as an energy source. A possible alternative would be to

TABLE 17 - COMPARABLE WATER ANALYSIS OF REFUSE POND ASH

DISPOSAL POND[17]

	Mississippi River	Refuse Pond Influent	Refuse Pond Effluent	Typical Bottom Ash Pond	Missouri Limit
Alkalinity (CaCO$_3$)	128	144	178	122	-
Aluminum	2.5	3.5	0.5	0.5	-
Barium	0.15	0.25	0.20	-	1.0
BOD (5 day)	4.0	6.0	0.0	4.0	50
Cadmium	<0.01	<0.01	<0.01	<0.01	<0.01
Chromium (hex)	<0.01	<0.01	<0.01	<0.01	0.05
Chromium (total)	<0.01	<0.01	<0.01	<0.01	1.00
Chromium (tri)	<0.01	<0.01	<0.01	<0.01	1.00
COD	26	190.5	0.0	0.0	150.0
Copper	0.04	0.04	0.01	0.01	0.02
Cyanide	0.003	0.002	0.002	0.002	0.02
Iron (total)	4.25	5.00	0.55	0.25	10.0
Kjeldahl N	1.12	3.36	3.58	1.68	-
Lead	0.04	0.02	0.01	0.01	0.10
Mercury (diss.)	<0.00005	0.00005	<0.00005	<0.00005	0.005
Nickel	0.12	0.04	0.01	0.01	0.80
Oil and Grease	1.2	1.52	2.71	0.016	15.0
pH	7.46	7.28	7.63	8.16	6.5 - 9.0
Selenium	0.04	0.06	0.03	0.03	0.01
Silver	<0.01	<0.01	<0.01	<0.01	0.05
Solids (diss.)	232	276	370	347	1000+
Solids (settleable)	2.9	9.2	0.01	<0.01	0.1
Zinc	0.07	0.29	0.04	0.02	0.10

Note: Units of settleable solids is ml/l. All others are expressed as
        mg/l, except pH.

accept environmental trade-offs. That is, in eliminating one environ-

mentally unacceptable disposal method, we would be creating one of

lesser environmental impact but with many side benefits. Obviously

such a situation would only be possible under the auspices of governmental

regulatory agencies. The ultimate elimination of land blights such as

landfills and strip mines and energy problems may justify some air and

water pollution.

Economics

The actual value of solid waste burning is dependent on the cost
of conventional fuels, the actual heating content of the solid waste
and the cost of conventional disposal methods. Therefore, the gross
value of solid waste can vary from $9 to $2 for 10 million BTU per ton.[16]
Table 18 gives the projected costs for solid waste supplementary fuel
systems. These costs should be taken along with the fact that many
solid waste items have actual and/or potential markets for recycle.
The magnetic metals and fly ash are already being utilized.

The economics of a solid waste power plant would be based on
several factors including:

1) Governmental regulations of solid waste landfill sites.

2) Air and water pollution effected by the process.

3) Transportation, handling, and processing costs.

4) The cost effectiveness of alternate methods of solid waste
disposal and/or utilization.

5) Sanitary landfill costs.

6) The recovery of resources such as magnetic and non-magnetic
metals and their market values.

7) Convention fuel costs.

Aside from these, the ultimate value is to the community based
on less landfills and more energy from an inherently problem oriented
effluent.

TABLE 18 - PROJECTED COSTS FOR A SOLID WASTE

SUPPLEMENT FUEL SYSTEM[16]

	Smaller Systems (30 tons per hour)	Larger Systems (125 tons per hour)
Processing facilities+		
Capital cost, per ton of daily capacity	$3,500 to $4,500	$2,000 to $3,000
Capital cost per ton++		
Typical public financing	$1.40 to $1.80	$ .80 to $1.20
Typical private financing	$2.20 to $2.90	$1.30 to $1.90
Operating costs per ton	$4 to $6	$3.50 to $5.50
Transportation facilities, including amortization		
Simpler cases	$.50 to $1 per ton	
Complex cases	$5 to $6 per ton	
Firing facilities		
Capital cost, per ton of daily capacity	$3,000 to $3,500	$2,000 to $2,500
Operating costs, including amortization		
Favorable circumstances	$.50 to $1 per ton	
Less favorable circumstances	$2.50 to $5 per ton	

+ Basic parameters of the processing facilities: two-stage milling, with air
  classification after the first hammermill; two 8-hour shifts per day, 250
  operating days per year; land costs are not included; residue disposal cost
  is not included.

++Typical public financing reflects a 6 percent cost of capital over a 15-year
  life. Typical private financing reflects a 10 percent cost of capital over a
  10-year life. A shorter life is used in the private sector to assure the
  desired return on investment.

The Future St. Louis Solid Waste Utilization System

        In February of 1974, the Union Electric Company formulated plans

to develop a solid waste utilization complex capable of processing all

the solid waste generated by the entire metropolitan St. Louis region.

The system is scheduled to be on line by mid-1977.

800 tons of waste per day or 2.5-3 million tons per year will be handled by five to seven collection-transfer centers, owned and operated by handling the Union Electric Company. Residential as well as commercial and industrial wastes will be transported by rail to the waste processing facilities at the Meramec and Labadie Power Plants. The wastes will be packed in 100 cubic yard containers.

The Meramec plant presently consists of two 125-MW boilers, a 270-MW and a 300-MW front-fired, pulverized coal boiler. All four units are to be equipped to burn a composition of 20 percent processed waste and coal. Two thousand tons of raw waste per day is to be processed and burned at the plant.

The Labadie plant consists of four 600-MW tangentially fired, pulverized coal fired generating units. Steam conditions will be 1000°F at 2400psig. All four units will be equipped to burn a fuel composition similar to the Meramec plant. Just two units will be capable of utilizing the burnable fraction of the 6,000 tons of raw waste processed per day at the plant.

The raw wastes include domestic wastes, food scraps, cans, appliances, commercial wastes, wasted lumber, office refuse, and selected industrial solid and liquid wastes will be received at both power plants by rail car. A two stage hammermill will reduce particle size to one inch or smaller.

Air separation will follow and the combustible portion will then be transported pneumatically to the waste-firing burners in the boiler furnaces. Stand-by equipment will be available at each plant to keep downtimes at a minimum. The heavy non-combustibles will be

further separated into organic, glass and ferrous at the hammermills
for a further size reduction before disposal at landfill sites.
Conventional screening methods will effect glass removal after the
magnetics are removed, they will be densified in hammermills for sale
as No. 2 bundle scrap and the non-magnetic metals will also be sold
for scrap. Markets for recycled glass must be pursued to ensure a
complete resource recovery system. Although solid waste composition
varies, a project such as this is feasible from the technological as
well as economical viewpoint.

From Table 19 we note the refuse weight composition at the
St. Louis project is generally 80 percent burnable. With this fact
and the recovery of the other components in mind, we should rapidly
come to the realization that this refuse is a valuable resource. In
light of the success of the St. Louis project, perhaps in the future
we might consider a more descriptive term for these materials - Salvage
Worths.

TABLE 19 - COMPOSITION OF TYPICAL ST. LOUIS REFUSE

	Percent by Weight
Burnable	80
Glass	11
Magnetic Metals	7
Non-magnetic Metals	1
Rock, gravel, etc.	1
	100

CHAPTER 4

THE EAST BRIDGEWATER MASSACHUSETTS ECO-FUEL^{TM} II

RESOURCE RECOVERY PLANT

After examining the difficulties associated with present solid waste recovery systems, a joint venture by Combustion Equipment Associates and Arthur D. Little effected The Eco-Fuel^{TM} II Resource Recovery Plant. Their study showed some of the existing problems with resource recovery to include:

1) High investment and operating costs of shredder equipment.

2) Short shredder hammer life.

3) Low recovery efficiencies and high ash contents of the fuel are a result of poor classification systems.

4) High moisture content of the municipal solid waste results in poor waste classification.

5) Inefficient energy recovery from the municipal solid wastes.

6) Storage difficulties because of the high moisture content of the fuel fraction.

7) Poor handling properties and low bulk densities of the fuel result in high handling, storage, and transportation costs.

8) High moisture and ash content result in a fuel with a low heating value.

9) High maintenance costs to fuel feed lines effected by glass and dirt abrasives.

10) Poor combustion qualities of the fuel because of high moisture content and large particle size.

11) High particulate emissions because of high fuel ash contents.

12) Incomplete combustion results in the need to dispose of organics along with boiler bottom ash.[19]

The initial process, the Eco-Fuel[TM] I, was developed to minimize these problems by utilizing low-energy input shredding, classifying, and drying systems to generate a fuel of 11.5% ash and 10% moisture contents with a heating value of 6900 BTU's/lb. The resultant fuel from the Eco-Fuel[TM] I process had a low sulfur content and was odorless and stable over indefinite periods of time.   However, this fuel had a density which was only 7-10 lbs/ft^3. This density was too low to offer serious alternatives to the problems of solid fuel waste storage and transportation.[20] Further research and development, and a year and a half later resulted in a more effective process. This was the development of the Eco-Fuel[TM] II process; and a re-duction in ash content to 5 percent, moisture content to 2 percent, an increased density of the fuel to approximately 30 lbs/ft^3 and an increased heating value to 7500-8000 BTU's/lb.[19]

Eco-Fuel^{TM} II Process

The process flow diagram of the Eco-Fuel^{TM} II system is shown in Figure 13. Packer trucks deliver the municipal solid waste to an enclosed building with a slanting floor for storage before it is fed into the recycling system. The enclosed storage area effects odor control and the prevention of solid waste scattering, i.e. blowing of papers. The storage floor was designed at an angle for ease of operation of the solid waste, specially equipped, front-end loaders which feed the recycling system.[21] A conveyor belt transports the solid waste from the storage area to the primary shredder (a flail mill) where a size reduction takes place to produce a more uniform and manageable particle. The shredded waste passes to a magnetic separator which removes the ferrous metals, approximately 6-8 percent by weight. The shredded waste that remains then passes on to a screening operation which classifies the waste into three fractions: 1) those particles which are relatively large and are returned to the primary shredder, 2) those particles which are relatively small, principally glass and dirt, 3) the solid waste fraction that will be further treated. The solid waste fraction is conveyed to a chemical treater and reduced to a fine particle size. Exact details as to the process operation are not available because patents are pending. However, the manufacturer's claims show a relatively low energy requirement (10-30 HP/ton).

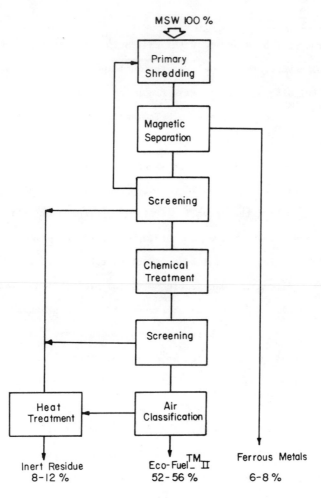

FIG. 13.   The Eco-Fuel^{TM} II process flow diagram.[19]

The chemical treatment process utilizes catalysts which are
recycled, to aid in size reduction, but do not alter the chemical
composition of the solid waste. Although it is not specified, it
is assumed the Eco-Fuel^{TM} II process, as do most chemical size

reduction processes, utilizes liquid chemicals which require dryers
to reduce the moisture content prior to the secondary screening and
air classification. The secondary screening and air classification
further reduce the ash content of the Eco-Fuel^TM II (the final fuel
product of the process). The glass, dirt, ash and coarse materials
from the screening and classifying steps are heat treated and
sterilized to produce a residue suitable as clean fill. This
residue can also be further separated for the recovery of glass and
non-ferrous metals.

Problems caused by plastics in other systems are solved by the
Eco-Fuel^TM II process by either removing the plastics or reducing
their size with the remainder of the organics and are completely
combustible.

The Eco-Fuel^TM II is the organic fraction of the processed
municipal solid waste and is an easily handled free flowing fine
powder. Table 20 gives some of its properties. Particle size can

TABLE 20 - ECO-FUEL^TM II PROPERTIES[19]

Particle size:	1/4" to -100 mesh
Heating value:	7500 -10000 BTU's/lb
Moisture content:	Less than 2% by weight
Inorganic content:	Approximately 5% by weight
Storage Life:	Indefinite
Bulk density:	Approximately 30 lbs/ft^3
Handling properties:	Free flowing powder

FIG. 14. Pictorial description of Eco-Fuel™ II.

A. Artist's rendition of a Combustion Equipment Associates Resource Recovery plant.

B. City of Brockton trucks discharge refuse at East Bridgewater, Massachusetts, Resource Recovery plant.

C. Municipal refuse is received and stored on tipping floor before being fed to recycle system by a specially equipped front-end loader.

D. Raw refuse is conveyor-fed to primary shredder-classifier where the light combustible fraction is separated from the heavier non-combustibles.

E. Glass, metal shavings, and inerts are discharged from classifier.

F. Heavy waste fraction moves to metal separation, while light combustible fraction is conveyed overhead to initial screening process. Eco-Fuel™ II feedstock is then dried, ground, and further screened into final fuel product.

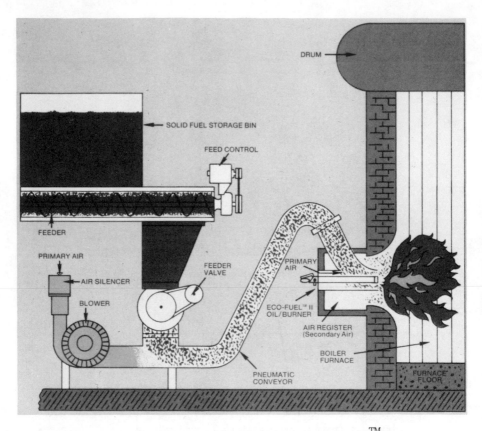

FIG. 15. A sustained commercial firing of Eco-Fuel[TM] II, CEA's proprietary fuel. The flow schematic for the firing is shown. The fuel is removed by a controlled screw-feed from a storage bin, at the rate necessary to maintain required boiler output. The fuel then passes into a pneumatic air transport system where it is carried to a standard combination oil/ pulverized coal burner.

be varied to meet specific firing and therefore the Eco-Fuel[TM] II

can be used in a number of manners such as direct firing in boilers,

slurring in oil or waste fuels, or pyrolysis. The small particle

size of Eco-Fuel[TM] II makes it ideal for pyrolysis operations.

The small size makes handling and feeding easier, increases exposed particle surface area to effect high heat and mass transfer and good control over process variables (time and temperature). The pyrolysis gas product can be used directly as a fuel; through upgrading to a higher heat content fuel or commodity chemical, its potential uses can include:

1) Boilers or gas turbines fuel.

2) Direct addition to a pipeline.

3) A methanol, ammonia, or formaldehyde feedstock.

4) An iron and steel reducing gas.

5) A quality gas feedstock.

6) A hydrogen feedstock when it is manufactured as a fuel for fuel cells.

The East Bridgewater Massachusetts Eco-Fuel^{TM} II System

The East Bridgewater Massachusetts Eco-Fuel^{TM} II plant can process 1200 tons per day[20] of municipal solid waste generated by the communities within a 10-25 mile radius of East Bridgewater. Each city with a contract pays $9.25 a ton to the Eco-Fuel^{TM} II plant to accept and process all of their solid wastes. The Eco-Fuel^{TM} II is then generated and sold to the power generating utility on a price per BTU delivered basis.[22] This allows for variations in solid waste BTU/lb quality. Revenues from the sale of the solid waste fuel and the sale of magnetic metals produce an economically competitive system for Combustion Equipment Associates.

The Landgard Resource Recovery System was developed by Monsanto Enviro-Chem Systems, Inc. of St. Louis. They initially operated a 35 ton per day pilot plant in St. Louis County, Missouri which demonstrated the technical and economical feasibility of energy recovery from unclassified municipal solid wastes. This pilot plant was in operation through 1971. Operation of this project further enabled Monsanto to study and demonstrate the process' feasibility:

1) To demonstrate the continuous and direct fire pyrolysis of typical unclassified municipal solid wastes.

2) To gather data and experience to design the most efficient scaled up facility.

3) To gather data to ensure prospective buyers of the system's performance.

4) To develop and improve the residue resource recovery system.

5) To demonstrate the facility to governmental agencies.

The success of this pilot plant has resulted in the City of Baltimore building a 1000 ton per day mixed municipal solid waste processing plant. The plant will be able to handle large appliances and tires.

The $16 million project is funded partially by an EPA grant ($6 million), by a Maryland Environmental Service (MES loan - $4 million), and a contribution by the City of Baltimore ($6 million). Revenues obtained from the recovered resources will be used to pay back the MES loan.[23]

The Landgard Resource Recovery System encompasses all phases of the recovery operation.

Phase 1 - Receiving, handling, shredding and pyrolyzing the waste.

Phase 2 - Quenching and separating the residue.

Phase 3 - Generation of steam from waste generated heat.

Phase 4 - Purification of the off-gases.

The basic pyrolysis process is the physical and chemical decomposition of organic matter effected by heat in an oxygen deficient atmosphere. The pyrolysis action on the solid waste generates gases with relatively high heating values. The solid waste has a weight reduction of 75 percent and a volume reduction of 95 percent.[23-26] The product remaining is a residue composed of metals, glassy aggregate and char. The metals and glassy aggregate can probably be recycled leaving only the odor-free and most inert char to be disposed of by landfill.

     A process flow diagram is given in Figure 16 , and
Figure 17 is a block chart of the same process. The Landgard
System is shown to be a complete energy producing facility with
receiving, processing and disposal facilities on site.

The Landgard Process

    The Landgard Resource Recovery Center receives raw solid
waste from conventional collection trucks six days a week. The
refuse is discharged into a concrete pit and bulldozed onto conveyors to be passed to the two shredders. The two hammermill
shredders can handle large and bulky items, refrigerators, mattresses,
and tires, as well as other solid waste items and reduce them to
uniformly sized particles (4 inches). The shredding process also
reduces odors and makes handling the waste easier. The receiving

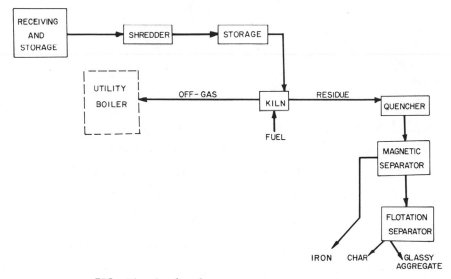

FIG. 16. Landgard resource recovery system.

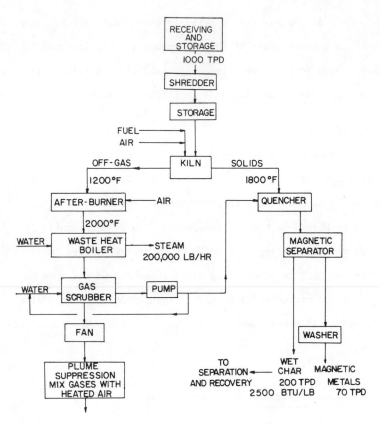

FIG. 17. Process flow block chart of Landgard resource
recovery system.[25]

and shredding processes are housed in ventilated buildings for weather

protection and for odor, noise, and shredded waste scattering control.

The receiving pit has a 1000 ton capacity and the hammermill shredders

are driven by 900 Hp motors.

The milled waste is transported to a live bottom storage bin

which constantly feeds into the pyrolysis reactor. The 2000 ton

capacity allows for continuous operation -- 24 hours a day, 7 days

a week. The pyrolysis reactor is a specially designed refractory
lined rotary kiln, 19 feet in diameter, 100 feet long and weighing
134 tons. Shredded waste enters one end of the kiln, and No.2 heating
oil the other end. To achieve the desired heat output 7.1 gallons
of fuel oil are combusted with every ton of solid waste. Rotation of
the kiln exposes more surface area of the solid particles to the high
temperatures and allows for uniform combustion. About 40 percent
stoichiometric air is added to aid combustion and maintain the pyrolytic
reaction. Pyrolytic gases have a countercurrent movement to the solid
waste exposing it to progressively higher temperatures. The gases
exit at the solid waste entrance to the kiln. As a result of the
initial contact with gases that are controlled to 1200°F, moisture is
removed from the solid waste material before pyrolysis occurs. The
waste residue must be kept below 2000°F to prevent the glass particles
from melting and creating a giant mass. The pyrolytic reaction is
optimized by this process design and was achieved through testing
and data gathering at the Monsanto pilot plant.[18,23,24,27]

Upon leaving the kiln, the hot residue is discharged into a water-
filled quench tank where a conveyor dewaters the wet residue and
elevates it into a flotation separator. In the separator the light
materials float off as a carbon char slurry which is passed through a
thickening and screening process to effect water removal. This water
is recycled to the closed loop water system of the plant. The wet
char is conveyed to a storage pile and then trucked to a landfill
site. The sink fraction or heavy material is conveyed from the bottom
of the flotation separator and passed through a magnetic separator

for magnetic metal removal. The magnetic materials are deposited in a
storage area and/or directly into shipping containers to be sold as
scrap. Nearly 65 percent of the sink fraction is glass (glassy
aggregate) which passes through a screening operation for size
separation and then stored on-site to be used in the manufacture of
"glass phalt" to pave streets. Pyrolysis gases are withdrawn from the
kiln to enter a refractory-lined gas purifier (afterburner) and mixed
with air and combusted. The afterburner combustion temperature is
approximately 1800°F which not only ensures complete combustion of the
gases, but also destroys odors. Table 21 shows the contents of these
gases which have a heating value of about 120 BTU/dry cubic ft.

The hot combustion gases then pass through water-tube-boiler
heat exchangers to generate steam at the rate of 200,000 pounds per
hour. The exiting gases are cleaned and cooled of particulates by
a water spray gas scrubbing tower and dehumidifier. An induced-draft
fan pulls the gases through the entire system and pushes them out the
stack. Table 22 shows an analysis of this gas as it leaves the stack.
Emissions are generally below local air emission regulations. Removed
particulate matter is recycled to the process thickener to aid solids
removal. The scrubber system requires little makeup water which is
recycled in the closed loop system.

The Baltimore pyrolysis process generates low BTU gases with low
particulate levels making them attractive to be combusted along with
coal, oil or natural gas in utility, industrial or institutional
boilers which are in close proximity to the plant.

TABLE 21 - ANALYSIS OF LANDGARD PYROLYTIC COMBUSTION GASES[23,25]

Nitrogen	69.3
Carbon Dioxide	11.4
Carbon Monoxide	6.6
Hydrogen	6.6
Methane	2.8
Ethylene	1.7
Oxygen	1.6

Solids Content	1-2 Gr/DSCF
Moisture Content	19% by volume

Heat Available	250 $\bar{M}$ BTU/hr (HHV)
(from 1000 TPD)	50 $\bar{M}$ BTU/hr (sensible)
	300 $\bar{M}$ BTU/hr

A flow diagram of this process is provided in Figure 18. The pyrolytic gases are withdrawn from the kiln, bypass the afterburner and are fed directly into the utility boiler.

The Baltimore Maryland Langard System

The Baltimore system can handle up to 1000 tons of mixed municipal solid waste per day. After pyrolysis, the plant will yield approximately 70 tons of ferrous materials which are under contract to

TABLE 22 - ANALYSIS OF PYROLYTIC EXITING STACK GASES[25]

STACK GAS ANALYSIS* Volume %	
Nitrogen	78.7%
Carbon Dioxide	13.8%
Water Vapor	1.8%
Oxygen	5.7%
Hydrocarbons	10 PPM
Sulfur Dioxide	150 PPM
Nitrogen Oxides	65 PPM
Chlorides	25 PPM
Particulates (NAPCA method)	0.02 grains/scf dry gas corr. to 12% CO_2

*Emissions are lower than conventional boiler systems.

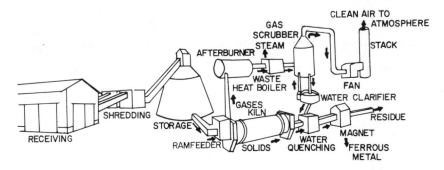

FIG. 18. The process flow diagram of the Landgard resource recovery system.[23]

be recycled and 170 tons of the glassy aggregate which can be used as

paving material for city streets or as clean landfill. The glassy

aggregate is expected to have a value of $2 per ton. The carbon char

residue, about 80 tons, or 6 percent of the plant throughput must be

disposed of through landfills. The residue is odorless and noncontamina-

ting and approximately 0.1 percent putrescible material and is available

for land reclamation. In the future, this residue may also be utilized

in sewage effluent treatment. Tables 23-25 are analyses of the various

pyrolysis residues.

FIG. 19.   Baltimore Landgard 1000 TPD plant.   (Monsanto Co.)

FIG. 20. Baltimore Landgard, gas scrubber, fire end of rotary
kiln including water quench tank and residue conveyor. (Monsanto
Co.)

4.8 million pounds of steam per day is recoverable from this

pyrolysis system. A 4500 ft Baltimore (BG&E) Gas and Electric Co.

steam main transports the steam to the distribution lines to be

utilized in the city's downtown area for heating and cooling. Costs

for the steam vary with the cost of No. 6 fuel oil and payment is

made by BG&E on 1000 pounds of steam basis.

FIG. 21.   Baltimore Landgard-2000 TPD Atlas storage unit.
(Monsanto Co.)

In order to maintain its 1800°F operating temperature, the
kiln must burn 2.2 million gallons of fuel oil per year. However,
in order to generate a comparable amount of steam through conventional
methods 18.2 million gallons of fuel oil must be burned. The result
is a net saving of 16 million gallons annually.[24]   A general cost
study of the Baltimore Plant is shown in Table 26. Note that as the
price of oil keeps rising, the system's economics look more and
more promising.

FIG. 22. Baltimore Landgard weigh scale, shredder building, gas scrubber. (Monsanto Co.)

TABLE 23 - ANALYSIS OF CARBON CHAR RESIDUE[23]

Bulk Density	20-50 pounds per cubic foot
Moisture Content	50% by weight
Heating Value, dry basis	7,000 BTU per pound

Analysis, dry basis

Component	Percent
Carbon	50.0
Ash and Glass	45.8
Volatiles	4.0
Sulfur	0.2

Analysis of water-extractable fraction

Component	Percent or parts per million (ppm)
Sodium	over 30%
Calcium	0.1 - 1.0%
Copper	0.03 - 0.3%
Magnesium	0.03 - 0.3%
Potassium	0.03 - 0.3%
Boron	0.01 - 0.1%
Strontium	0.001 - 0.1%
Iron	0.001%*
Molybdenum	0.001%*
Silicon	0.001%*
Phosphorus	25 ppm*
Chromium	10 ppm*
Lead	10 ppm*
Tin	10 ppm*
Vanadium	5 ppm*
Zinc	5 ppm*
Aluminum	1 ppm*
Cadmium	1 ppm*
Manganese	1 ppm*
Silver	1 ppm*
Titanium	1 ppm*

*Less than figure shown.

TABLE 24 - ANALYSIS OF RECOVERED FERROUS MATERIALS[23]

Quality of Ferrous Metal Recovered from Pyrolysis Residue	
Bulk density	35 pounds per cubic foot
Iron	98.85% by weight
Contaminants	1.15% by weight

Chemical analysis

Component	Percent	Component	Percent
Iron	98.850	Antimony	.020*
Tin	.153	Sulfur	.016
Carbon	.150	Phosphorus	.015
Copper	.150	Cobalt	.010*
Nickel	.140	Molybdenum	.010*
Lead	.088	Titanium	.010*
Manganese	.048	Vanadium	.010*
Silicon	.045	Aluminum	.001*
Chromium	.035	Other	.249

*Less than percent shown.

TABLE 25 - ANALYSIS OF RECOVERED GLASSY AGGREGATE[23]

Bulk Density	150 Pounds Per Cubic Foot
Component	Percent
Glass	65
Rock and miscellaneous	28
Ferrous metal	3
Nonferrous metal	2
Carbon	2

TABLE 26 - ECONOMIC EVALUATION OF THE BALTIMORE PYROLYSIS PLANT
($ per throughput ton)

Costs and Revenues	January 1973	February 1974
Amortization*	$4.34	$5.55
Operating Costs		
Fuel	$ .89	$2.20
Electricity	1.06	1.50
Manpower	1.02	1.10
Water and chemicals	.31	.30
Maintenance	1.84	1.90
Miscellaneous	.42	.40
Char removal	.18	.20
Total	$5.72	$7.60
Total Expenses	$10.06	$13.15
Revenues		
Steam+	$3.89	$11.18
Iron	.44	1.55
Glassy aggregate	.34	.40
Total Revenues	$4.67	$13.13
NET OPERATING COST	$5.39	$ .02

*Approximate plant cost: in January 1973, $16 million; in February 1974,
$20 million.

+Price is keyed to fuel oil price, which was $3.70 per barrel in January 1973, and
$10.63 per barrel in February 1974.

CHAPTER 6

REFUSE TO ENERGY - SAUGUS, MASSACHUSETTS

A 1200 ton-per-day solid waste burning, steam generating plant is being constructed at Saugus, Massachusetts. It will service 16 communities of the north Boston area for refuse disposal and provide steam to a nearby industrial plant for processing and power generation. This solid waste energy process burns refuse without the need for supplemental fuels and meets strict environmental regulations for odors, noise, and air and water emissions. Initially the combustion residues will be used for road fill, and metals will be recovered. However, more extensive materials recovery processes will be developed as markets for these materials become technically and economically available. Future plant capacity can be increased to 2400 tons per day if the present facility, scheduled for completion at the time of this writing, proves successful.

The plant is expected to generate 2 billion pounds of steam annually which will be sold to the General Electric Company manufacturing plant at Lynn, Massachusetts across the Saugus River. Revenues from the steam will reduce disposal charges to the communities, each

of which is responsible for its own solid waste collection and transportation to the refuse plant. They will be charged a tonnage fee for
disposal. Furthermore, not only will this process save nearly 70,000
gallons of fuel oil per day, but sulfur and particulate emissions will
be lower than the replaced facilities burning low sulfur fuel oil.[28]

The Saugus, Massachusetts Project is unique in that it was
forced upon the 500,000 inhabitants it serves. A court order closed
down their sanitary landfill operations for environmental reasons.
A concerted effort by the communities, the General Electric Company,
and the M. DeMatteo Construction Company (owner's of the landfill
operation) was made to examine environmentally sound and economically
feasible disposal systems.

The result was the formation of the Refuse Energy Systems
Company (RESCO). Wheelabrator-Frye, Inc. and M. DeMatteo Construction
Company jointly own RESCO which in turn will own, operate, and finance
the new facility. Deliveries of steam began in the fall of 1975 to
the General Electric plant, and deliveries are to be continuous under
a long-term contract. Present boilers at the GE plant are to be
retired.

The solid waste plant is a private operation paying real
estate and income taxes. The initial capital investment of $30 million
takes into account several conditions not normally considered by a
solid waste disposal plant including:

1)   The plant must meet strict and enforced environmental regulations.

2)   Reliability of the process is essential because:

    a)   No alternate disposal methods are available.

    b)   Steam deliveries to GE must be continuous.

3)   To effect this reliability an extraordinarily large storage pit is necessary for periods of plant shutdown. An alternate oil feed system to the solid waste boilers and oil fired auxiliary boilers are planned to meet steam demand during shutdowns.

4)   The plant is built on the present landfill site requiring hundreds of piles to be driven 80 feet down into bedrock and a heavy concrete foundation.

5)   A single collection system will be in use to cut collection costs, but this has resulted in an additional sorting process and increased costs to the plant's construction bill.

6)   To convey the steam to the GE plant, construction of a utility bridge and half a mile of pipeline across the Saugus River is required.

Figure 23 shows the operation of the facility.

The RESCO Process

        The basic plant requirements are to process 1,200 tons per day

of domestic and commercial solid wastes and to generate steam with

properties of 625 psig and 787-825°F, continuously, 7 days a week, to

be delivered to the GE plant. Steam is expected to average about

300,000 pounds per hour and vary between 350,000 and 65,000 pounds per

hour. Costs of the steam will be less than conventionally generated

steam, but the charges will fluctuate with the cost of oil.[28]

        Figure 23 shows the process flow plan. Upon arrival at

the plant the collection trucks are weighed on the scales located at

the plant entrance. The solid wastes are then dumped into the

solid waste storage pit. The trucks are then weighed as they exit

the plant to figure the weight differential or the amount of wastes

dumped and the charges.

        The storage pit will have a normal capacity of 2,700 tons

or two and three days' firing potential. This capacity can be

increased to 6,700 tons or 5.6 days' operation and will be necessary

for future plant expansions.

        All open collection trucks containing mostly bulky materials

will be directed to the area of the refuse pit near the shredder. See

Figure 24. Two overhead cranes will be in operation. One crane

will load bulky materials which must be ground into the shredder.

These shredded materials are to be recycled back to the refuse pit where

the second crane loads new refuse to the incinerator/furnace feed hopper.

Refuse loaded into the incinerator feedhopper will not only fire the

furnaces but will act as a seal to prevent flame flashbacks and to

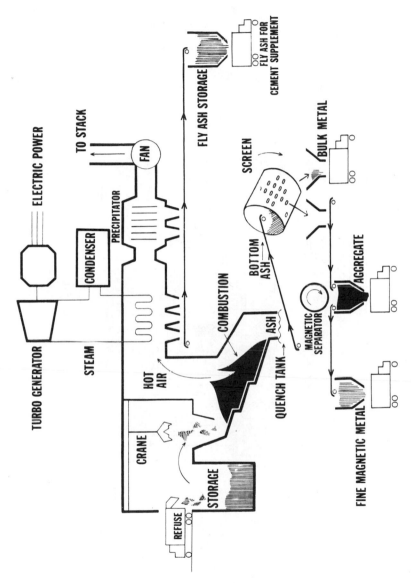

FIG. 23. Wheelabrator refuse disposal and resource recovery system.

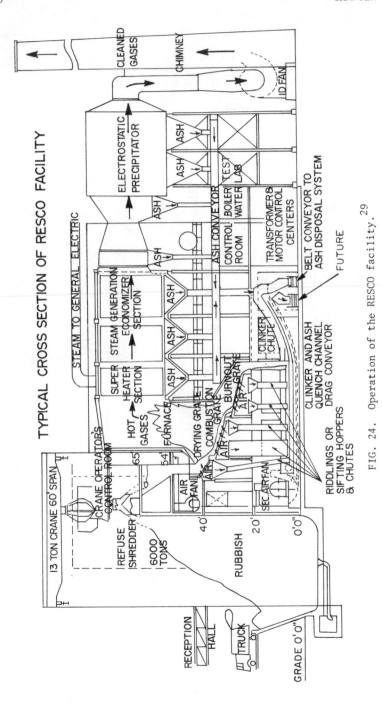

FIG. 24. Operation of the RESCO facility.[29]

maintain furnace draft. The height of the refuse in the hopper is to
be monitored to ensure this seal is maintained. The refuse then is
discharged through the bottom of the feedhopper onto the first of a
series of reciprocating grates.   The back-and-forth movement of the
grates helps to expose more surface area of the solid waste to the
heat of the furnace. The refuse moves down (see Figure 25) the first
grate, is dried, combustion begins and is then dumped onto the second
grate. Combustion is nearly completed on the second grate, and then
the combusted wastes drop onto the third grate where the final burnout
occurs and only ash remains. There will be two 750 ton-per-day
incinerators, each with its own feed system.

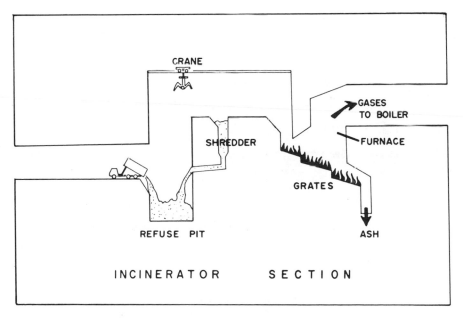

FIG. 25. Process flow of the incinerator/furnace.[29]

The ash generated by the combustion process will be collected in hoppers and passed through a water filled channel for quenching. An alternate quench channel will be available to ensure plant operations when the channel must be shut down for maintenance purposes. After quenching, the ash will pass through a screening process and magnetic metals separator. The magnetic metals will be sold for scrap and other materials economically recoverable will also be sold. The remaining ash will be trucked to landfill sites as inert landfill.

Primary combustion air will be ducted to the underside of the gratings to effect burning of the refuse. It will be drawn from the refuse storage pit building as an odor control measure. Odors from the refuse pit will be destroyed in the incinerator. Steam coil air heaters may be installed in the ductwork should heating of the combustion air become necessary. Secondary combustion air will be blown over the burning refuse from either side of the grates. Its purpose will be to further the combustion process and reduce the flue gas temperature. The flue gas temperature must be controlled to between 1600 and 2000°F to inhibit corrosion and inhibit the formation of thermal NO_x (nitrogen oxides formed as a result of the oxidation of combustion air nitrogen).

The flue gases leave the furnace/incinerator and pass through the superheater, generator section and economizer and then exit the boiler. Of the 300,000 pounds-per-hour of steam generated by the boilers about 5 percent is required by the RESCO solid waste facilities for plant operations and the rest is conveyed to the GE plant.

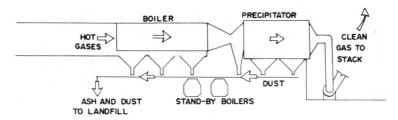

FIG. 26.   Process flow of the boiler-precipitator section.[26]

Pollution Control at RESCO

The storage pits are to be enclosed to inhibit odor emissions
by refuse. As previously noted, air will be drawn from the storage
pit area to be used in the combustion process. This will create a
negative pressure in the storage pit building; air will be drawn into
the building through its truck entrance and very small amounts of
odorous emissions are expected to leak from the structure. Odors from
the refuse pit will be destroyed in the furnace/incinerator.

To reduce fly ash and other particulate emissions below governmental regulations, each incinerator/furnace will be fitted with an
electrostatic precipitator (ESP's). Each ESP will handle 240,000 CFM
of flue gas at temperatures in excess of 420°F with dust loadings of
1 to 2 grains/SCF. The efficiency of each ESP is rated at 97.5 percent.
An induced draft fan will pull the gases through the precipitator and
push them out a common stack. Since the sulfur content of solid wastes
is relatively low, there are no problems with sulfur emissions. The
stack will be nearly 180 feet high and it will disperse gases consisting
almost entirely of nitrogen, carbon dioxide and water vapor.

FIG. 27. RESCO north shore facility resource recovery plant at Saugus, Massachusetts, showing location east side of Salem Turnpike which will provide easy access for refuse haulers. The plant is shown under construction in close proximity to a residential area which will not be burdened with any air or water pollution from the plant. Air pollution control will be achieved by two Wheelabrator Lurgi electrostatic precipitators. Designed and fabricated in the U. S., these units will meet or exceed all existing federal, state, and local governmental standards.

Municipal water will be used for process water and to a large extent will be recycled. Steam from the GE plant will return to the solid waste plant via the utility bridge; it will be treated and reinjected into the system. Wastewater generated by the ash quenching

process will be discharged to the Saugus city sewage system. No
waters will be withdrawn or discharged into the Saugus River.[28,29]

Benefits to the Community

The RESCO facility will alleviate many solid waste disposal
problems in the Boston area. It will be the nation's first privately
financed refuse recovery plant. The technology employed has been
utilized to a great extent in Europe, Japan, and Canada. The plant
concept was selected for its reliability for continuous operation,
pollution control, and steam generation.

The new RESCO refuse-to-energy conversion plant at Saugus,
Massachusetts will:

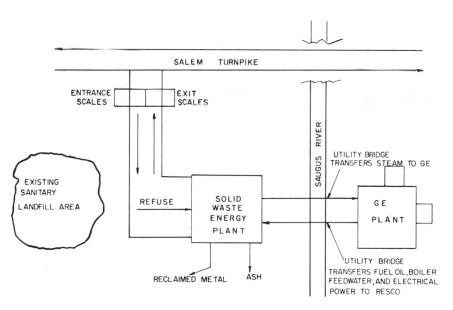

FIG. 28. Process flow of the Saugus, Massachusetts solid
waste power plant.

1) Dispose of 1,200 tons of garbage each day, about four pounds per person from a half-million people in 16 communities, equivalent to an acre of garbage piled three feet high.

2) Maintain the highest environmental standards.

3) Generate energy in the form of steam power -- to be converted to electric power for the nearby General Electric Company plant. This will help pay the disposal costs.

4) Conserve the shortening supply of fuel oil, which is currently being used to power the nearby plant.

5) Reclaim useable materials such as metal from the refuse during the process.

6) Produce only sterile compact residue -- one-tenth the original refuse volume.

7) Establish the North Shore communities as leaders in this field of clean refuse-energy recovery.

Summary Facts About the RESCO Plant

Refuse Energy Systems Company (RESCO), the owner and operator of the plant, is a joint venture of Wheelabrator-Frye, Inc. and the M. De Matteo Construction Company. The capital cost of the RESCO facility will be approximately $31 million. No governmental monies are involved in this project. The financing for the entire facility is being provided by Wheelabrator Financial Corporation. Because the plant is privately owned, it will be a taxpayer. When operational, the plant will employ approximately 50 people. The plant is being built in Saugus, Massachusetts, on the east side of

the Salem Turnpike, adjacent to the Saugus River. The plant site is
a 13-acre tract, reclaimed from a landfill area, and will process
1200 tons of municipal and commercial refuse per day. Refuse will be
trucked to the plant from 16-18 communities in the greater Boston area
(approximately 500,000 population). Present plans call for the recovery of resources, heat energy and ferrous metals. Design provisions
will allow the future recovery of additional materials.

RESCO has a 14-year contract with General Electric to supply
a minimum of 2 billion pounds of steam per year to its Lynn River Works
plant across the Saugus River. Steam from the RESCO plant will save
G.E. the equivalent of 73,000 gallons of fuel oil per day, by replacing the energy-producing capacity of two oil-fired power boilers
which will be retired by G.E.   The price of the steam produced by
the RESCO plant will vary according to the fuel price market. The
RESCO plant is expected to recover approximately 10 tons of ferrous
metals per day. The volume of sterile ash produced will be less than
10 percent of the volume of the refuse fuel.

Refuse will be fed into the plant's furnaces just as it
arrives at the plant. No separation or shredding is necessary.
Large, bulky items such as furniture and appliances will be fragmentized
before entry into the furnaces. (Ferrous metals will be recovered from
the ash). The plant's combustion temperature will be 1600° - 1800°F.
Steam pressure will be in the 690 psi range. The refuse will be
self-combusting. No auxiliary fuel source is needed. The plant's twin
furnaces are of the reciprocating-grate design, supplied by Von Roll Ltd.
of Switzerland. Von Roll Ltd. has 108 worldwide plants (211 individual

systems) operating or on order. The first installation at Berne,
Switzerland has operated successfully for 20 years. The boilers are of
Von Roll design, built by Dominion Bridge Co. of Canada.

Air pollution control will be achieved by two Wheelabrator
Lurgi electrostatic precipitators. Designed and fabricated in the
U.S., these units will meet or exceed all existing federal, state and
local governmental standards.   Rust Engineering Company (a wholly owned
subsidiary of Wheelabrator-Frye, Inc.) is responsible for design, engineering and construction management of the project. The M. De Matteo
Construction Company is the general contractor.   Construction of this
facility calls for 1,150 tons of structural steel, 1,250 tons of re-
inforcing steel, 14,000 cubic yards of concrete and over 600 pilings
driven 80 feet to bedrock. The plant design includes provision for
future expansion from the present capacity of 1,200 tons per day to
2,400 tons per day.

EUROPEAN STEAM PRODUCING INCINERATORS

The disposal of solid waste through incineration rather than through landfilling or composting is a necessity in Western Europe because a large solid waste volume reduction must be obtained in order to optimize the short supply of landfill and dumping sites. The burning of refuse, which may have an ash content of up to 60 percent, effects high particulate emission rates. These emissions require high-efficiency particle collection devices such as electrostatic precipitators. In order to reduce maintenance of the ESP's the gases entering the precipitators must be reduced below 600°F. European technology has developed a process to reduce these gas temperatures without utilizing wet scrubbers with high operating costs or combustion modifications which entail flue gas recirculation. The hot exiting incinerator gases would enter a steam generator which is used as a heat exchanger. The steam generated is a by-product of the solid waste disposal process and in actuality a benefit derived from pollution control. As a result of early experience in steam generation (even though at first it was only a by-product) there is widespread use of the incinerator/steam generator concept in Western Europe. Furthermore, this steam is readily available

for electric power production since many Euopean municipalities

are responsible for the generation of electricity as well as solid

waste disposal, reducing red tape and financing problems.

Thus we see many varied factors contributing to the widespread

use of incinerator plants, which burn over 120 tons-per-day, as steam

generators:

1)   The European population density.

2)   Where the cost of conventional fuels have risen in

the United States, prices in Europe are astronomical.

3)   District heating is an available demand in many European

towns for the steam output of these incinerator/generators.

4)   Through ownership of incinerators, electrical utilities,

and the district heating system, municipalities can plan the total

resource system.

5)   The cost of flue gas cooling and cleaning equipment is so

exorbitant that heat recovery through steam generation is essential.

Table 27 shows the typical European steam producing plants. Germany has

contributed much technology to solid waste incinerator/steam generation

and has more operating plants than any other country in the world.[30,31]

Germany has had refuse burning steam generators for many

years and much can be learned from their operating experiences, both

positive and negative. There are quite a number of different processes

in operation: Some units are designed for high steam pressures and

temperatures and others, relatively modest steam properties. Some
furnaces burn only solid waste, some burn solid waste as a supplement
to coal or oil and some burn the coal and oil as a supplement to the
solid wastes.

The type of process utilized is dependent on what the steam
will be used for. Initially, small incinerator plants produced steam
with relatively modest properties for district heating networks and
small amounts of electricity. However, since fuel shortages have
become so acute, the move has been to generate high pressure and
temperature steam: 1000-1500 psig and 750-925°F. A Munich plant
burning both fossil fuels and refuse produces steam with 2600 psig
and 850°F characteristics.

Figure 29 shows the process flow and the major components
of a typical modern European plant. It accepts a wide range of waste
items and generates products for widespread use.

Operating and Equipment Experience

Throughout Europe the process consists of a traveling crane
with a grapple feeding a stokes hopper with refuse. This produces a
fuel bed about 3 feet thick which is burned on the furnace grate firing
system. There are many different types of grate designs (some are
discussed below) but all are moving grates because it is believed
satisfactory combustion is obtainable only through continuous mixing
of the waste which produces a uniform bed with a constant tumbling
action. Some grate designs seem to have advantages over others;

TABLE 27 – TYPICAL EUROPEAN STEAM PRODUCING INCINERATION PLANTS[31]

City	Date Of Service	Stoker Mfgr.	Tons Of Refuse Per Day	Flow lb/hr x 1000	Pres. Psig	Temp.°F	Use*	Dust Separation
Germany								
Stuttgart	1965	VKW	530	110	1095	975	P	E
Darmstadt	1967	Von Roll	250	55	680	840	DH+P	E
Ludwigshafen	1967	Von Roll	250	50	600	805	DH+P	E
Frankfurt II	1967	Von Roll	400	53	910	930	P	E
Hamburg	1967	Martin	320	60	255	645	PA	-
Hamburg	1972	Martin	650	88	585	770	P	E
Nuremberg	1968	Von Roll	400	81	1195	840	P	E
Solingen	1969	Von Roll	270	55	680	840	P	E
Leverkusen	1969	Von Roll	270	62	285	580	DH	E
Munich II (a)	1966	Martin	1010	178	2560	995	DH+P	-
Munich III (a)	1969	Martin	1010	178	2560	995	DH+P	-
Munich IV (a)	1971	Martin	1010	178	2560	995	DH+P	E
Dusseldorf	1965	VKW	270	140	1280	930	P	-
Bremen	1969	VKW	330	33	365	420	DH	-
Iserloch	1970	Babcock	210	44	365	480	DH	-
Offenbach	1970	VKW	270	57	355	480	DH	E
Landshut	1971	Von Roll	220	15	285	660	DH	-
Berlin	1967	Duerr	330	49	1240	880	P	-
Berlin	1971	Duerr	420	49	1240	880	P	-
Berlin/GDR	-	VKW	400	77	1380	975	P	-
Switzerland								
Basel	1969	Von Roll	400	88	670	660	DH+P	E+ cycl.
Zurich	1969	Von Roll	290	62	525	790	PA	E
Luzern	1970	Von Roll	130	28	570	700	P	E
Hinwil	1970	Martin	130	27	555	750	P	-
Zurich	-	Martin	500	84	570	735	P	-

	Year							
Netherlands									
Den Haag	1968	Von Roll	400	78	570	800	P	E+ cycl.
Amsterdam	1969	Martin	600	95	610	770	P	-
Austria									
Vienna	1971	Martin	400	126	455	SAT	DH+P	E+ Cycl.
Sweden									
Stockholm	1968	Von Roll	130	(W)	230	250/160	DE	cyclones
Umea	1970	Von Roll	130	(W)	230	355/260	DH	cyclones
Goteborg	1970	Von Roll	400	104	285	420	DH	cyclones
Malmo	1969	Martin	230	1140(w)	130	Hot w.	DH	-
France									
Rennes	1968	Martin	130	35	385	440	DH	-
Paris	1969	Martin	130	295	1280	875	P	-
Strasbourg	-	Von Roll	390	72	780	860	DH+P	E
Metz	1970	Martin	160	40	215	635	P	-
England									
Edmonton	1971	VKW	370	89	625	850	P	-
Coventry	1972	Martin	320	70	250	405	P	-
Nottingham	1972	Martin	310	53	385	700	P	-
Soviet Union									
Moscow	1970	Martin	220	32	200	380	DH	-
Italy									
Genova	1971	Von Roll	220	42	465	705	P	E
Mailand	1971	Volund	270	50	455	825	DH	-
Finland									
Helsinki	1969	Von Roll	90	35	220	660	DH+P	-
Lahti	1969	Lokomo	130	26	570	840	DH	-
Norway									
Oslo	1966	Esslingen	170	44	355	500	P	-

* P = Power production   DH = District heating   E = Electrostatic precipitator
  a = Oil fired separately   PA = Plant auxiliaries   W = hot water

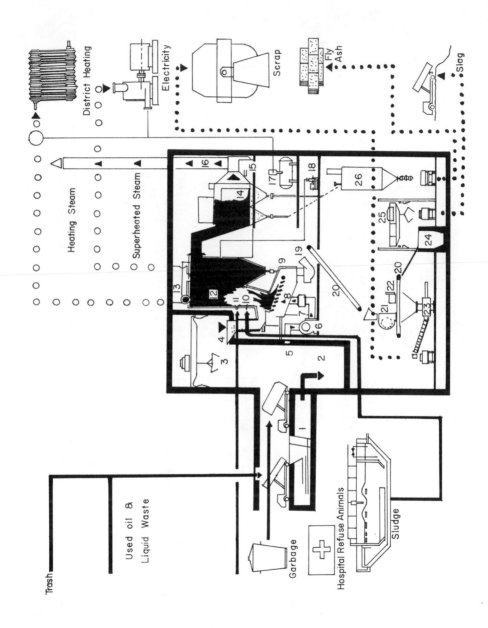

FIG. 29. Modern European plant.[31] 1, trash inlet; 2, garbage bunker; 3, crane; 4, garbage hopper; 5, air outlet from bunker; 6, under grate air; 7, air heater; 8, grate; 9, igniter; 10, sludge nozzles; 11, used oil burners; 12, boiler; 13, superheater; 14, electrostatic precipitator; 15, ID-fan; 16, slack; 17, deaerator; 18, feed water pumps; 19, slag outfeed; 20, conveyor; 21, pulverizer; 22, magnet; 23, scrap press; 24, slag bunker; 25, slag loading crane; 26, fly ash bunker.

however, most of the burning efficiency variations with the different
types of grates can be attributed to better fuel preparation (selective
shredding or controlled mixing) or improved control of the grates and
air supply to optimize fuel and air-flow characteristics. Much of the
concern with solid waste burning is equipment corrosion and the increased
maintenance costs. An analysis of European burning experiences and data
can reduce initial design problems and possibly remove the need for
costly pilot plants in America.

The emphasis in Germany is on a reliable continuous system
of refuse burning steam generators. A constant flow of waste through
a plant must be maintained to handle the solid waste disposal problem.
It is therefore common to have multiple steam generator units which
are used in rotation where one system is on line while the other is
down for maintenance. Solid waste flow can then be handled on a
continuous basis and the plant's availability (its ability to burn
refuse) is virtually 100 percent. This additional investment for plant
back-up equipment is necessary and is offset by the fuel savings
derived through burning solid wastes all the time rather than burning
coal or oil during periods when the system is down. Thus, relatively
small, but extremely high-steam pressure and temperature generating
steam units are utilized.

In general, the operation experience in Germany has found that
even if solid waste is burned as a supplement to fossil fuels in a
steam generating system, the resultant products of combustion act as
if they were the products of combustion of solid waste alone. This
proves to be costly since a very large steam generator with wide tube

spacings and low gas velocities is required for power production. This arrangement is necessary to prevent the collection of ash on the tubings.

Solid waste boilers are subject to the normal corrosive processes of conventional fuel boilers, and in addition unusual corrosive activity occurs. Some compounds liberated during solid waste combustion act as catalysts to increase corrosion. Another source of corrosion is the combustion of PVC effecting the formation of hydrochloric acid, HCl, which condenses on the boiler tubes.

1) If there is a deficiency of air at the tubes, the HCl reacts with the tube forming $FeCl_2$ which then evaporates.

2) Combustion gases react with steel tubing to form a protective oxide layer and $FeCl_2$.[31]

Excess air in the order of 50 to 100 percent above stochiometric and thorough flue gas mixing is necessary to prevent the above reactions. However, alkaline sulfates produced from solid waste combustion combine with the above reactions to effect highly corrosive activity on the furnace tubing.

These tubings are alternately exposed to oxidizing and reducing atmospheres even upon introduction to the furnace of 80 percent excess air. The corrosive activity effects the constant removal of iron oxide from the furnace side of the tubes. A refractory coating on the tubing inhibits corrosion and is the only available solution at this time. Atmospheres changing from oxidizing to reducing in the furnace are inherent in grate-burning systems and it was found that to ensure

oxidizing conditions at all times the solid waste burning process must
utilize 100 to 150 percent excess air.

When the furnace walls are covered with a refractory coating
to prevent rapid corrosive activity, gas temperatures to the superheater
and heat recovery equipment are increased as a result of the insulating
effect of the refractory coating. However, corrosion of the superheater
tubes takes place, rapidly at first, but stabilizes after the formation
of a thick oxide film, Fe_3O_4. Soot cleaning and conventional soot blowing
were unsatisfactory methods for cleaning the superheater tubes because
both methods removed the protective Fe_3O_4. This was the same problem
encountered in the furnace in which the changing atmosphere effected
the removal of the oxide film on the walls of the furnace tubing.
Water washing the tubes has proved to be the most effective method
for cleaning the superheater tubes.

There are three large German utility plants which burn
significant quantities of solid waste to generate electricity.
These are municipal electrical plants located at Mannheim North,
Munich North, and Stuttgart/Munster. Operating characteristics of the
three plants as well as the type of equipment used and plant output
are given in Table 28.

Generally, three designs of solid waste burning grates are
utilized in Europe. The first design is the multiple traveling grate.
This design consists of four grates traveling in series. The first
grate is the feeder which is inclined and located at the inlet of the
furnace. This inclined grate is followed by three horizontal grates
with most of the solid waste combustion taking place on the middle

TABLE 28 – SOME EUROPEAN PLANT OPERATING CHARACTERISTICS AND TYPES OF EQUIPMENT[30]

Plant	Munich North I	Munich North II	Mannheim-North	Stuttgart Munster	Stuttgart Munster
Number of Boilers	2	1	3	2	1
Electrical Output (Each), Mw	34	112	*	13	13
Pressure, S.H. Outlet, psia	2640	2640	1710	1095	1095
Temperature, S.H. Outlet, °F	1004	1004	977	977	977
Temperature, R.H. Outlet, °F	1004	1004	-	-	-
Type Grates	Gravity Flow Reverse Action	Gravity Flow-Reverse Action	Multiple Traveling Grates	Walzenrost "Multiple Barrels"	Gravity Flow Reverse Action
Boiler Manufacturer	German Bab-cock	German Babcock	KSG(EVT)	VKW	KSG (EVT)
Grate Manufacturer	Martin	Martin	KSG(EVT)	VKW German B&W	Martin
Refuse Burning Cap. (Each), U.S. Tons/Day	660	1060	480	530	530
Max. % Refuse (Design)	54	30	100	40	40
Max. % Cap. on Prime Fuel	100	100	100	100	100
Prime Fuel Type	Coal	Coal	Oil (Fuel Oil Capability)	Oil	Oil
Primary Steam Flow (Each), lb/hr	220,460	804,680	88,180	275,580	275,580

*Electrical Production & Urban Heating (Header System)

conveyor. The principal advantage of this grate design is that
the speed of any one or all of the grates can be modified to allow
for variations in refuse heating value. The Mannheim North plant uses
multiple traveling grates.

The second design is the multiple "barrel" or roller grate.
This design entails the use of tangentially mounted cylinder grates with
a common centerline at a slope of approximately 35 degrees to the
horizontal. The tumbling action exposes more solid waste surface
area for more efficient combustion. The grate surface temperature
is less than other grates because the rolling action of the cylinders
allows a relatively short contact time of grate and burning solid
waste. The Stuttgart-Munster plant uses this grate design.

The third design is called the Martin reverse-action gravity-
flow. The initial system design was based primarily on gravity flow
of the refuse.   The solid waste would move down the grate surface by
gravity. Newer designs have reverse action reciprocating grates which
move the refuse through the furnace.[30,31]

Names for each grate system design vary from plant to plant
and manufacturer to manufacturer; however, the basic design is the
same as noted when we compare the Martin reverse-action gravity-flow
design to the design in Figure 30 .

Depending on plant location, economics, and plant preferences,
processing of the sterile ash leaving the grates can be accomplished in
various ways. At Munich, Germany, the iron is separated from the
ash magnetically and densified into approximately one-foot cubes, and
sold for scrap. The remaining residue is trucked to a nearby landfill

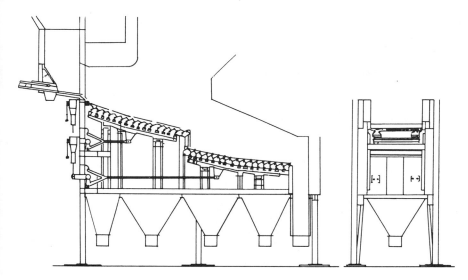

FIG. 30. Opposed motion revolving grate.

site. At Mannheim, Germany North, all' of the ash is removed by a
contractor daily, and ferrous metal separation and residue dumping are
carried out off-site.

The operation of many European incinerator/electric power
generating plants have refuse handling and corrosion problems that
would make incorporation of their processes into American utility
plants unfeasible. Although the Europeans have developed well-engineered
solutions to these problems, the basic process of raw refuse combustion
taking place on a moving grate would be unacceptable to an American
electrical power utility because the firing pattern of the fuel is
virtually uncontrolled. A given volume of raw refuse can have a heating
value of zero, negative, or relatively high, and refuse positioning
on the grate can vary from moment to moment. The principal advantage

of these processes is the production of a sterile ash from raw refuse.
Additionally, a relatively well-mixed solid waste has produced steam
in some units and maintained it to within $\pm$ 10 percent of steam requirements without requiring supplementary fuels. However, even with large
volumes of excess air, there are many occasions where a reducing
atmosphere exists in one part of the furnace or another. These localized reducing conditions can result in severe corrosion when corrosive
gases, low-melting chlorides, or sulfur-containing salts exist in the
flue gas.[30]

There is much technology to be learned from these processes.
Many problems can be averted by analyzing the experiences of these
operating plants, utilizing the available concepts and studying the
methods used to cope with the detrimental effects.

THE WORLD'S LARGEST INCINERATION PLANT IN FULL
OPERATION - NETHERLANDS

In the industrial section of Rotterdam, the Netherlands,
exists the world's largest solid waste incineration plant. Its
capacity is 700,000 tons of waste annually. The solid waste incinerator
plant is the major process of a $70 million municipal resource facility
which also includes an electric power station and a water distilling
plant.

The incinerator plant serves Rotterdam and 23 surrounding
towns and villages, and initially was to have an annual capacity of
190,000 tons of municipal solid wastes, 440,000 tons of industrial
wastes, and 70,000 tons of solid and liquid chemical waste products.
This facility went into operation in the Spring of 1973.

FIG. 31. Resource recovery plant, Frankfurt, Germany. In
operation since 1967, 1200 ton per day capacity, electrical
generation, steam for district heating. (Wheelabrator-Frye/Von Roll.)

General Background

In the mid-sixties the generation of solid wastes in the
Netherlands was accelerated as a result of population growth, an
increasingly industrialized socity, and the continued use of non-
returnable packaging. These factors were accentuated in the
Rijnmond Region, Rotterdam's New Waterway access to the North Sea.

FIG. 32. Resource recovery plant, Vienna, Austria. In operation since 1963, 600 ton per day capacity, electrical generation, steam for district heating and swimming pool, ferrous scrap recovery. (Wheelabrator-Frye/Von Roll.)

Rijnmond is the most densely populated region of the Netherlands with five times as many inhabitants per square kilometer as the rest of the country.

The generation of waste became so profuse that these wastes could not be processed in the existing refuse incineration plant in

FIG. 33. Resource recovery plant, Geneva, Switzerland. In operation since 1966, 400 ton per day capacity, electrical generation, hot water for industrial processing. (Wheelabrator-Frye/Von Roll.)

Rotterdam. Special landfill sites were allowed for the overflow. The portion that could be composted were transported to the northeastern part of Holland by rail and stored.

The available landfill sites began to diminish and environmentalists objected to the dumps on hygienic as well as

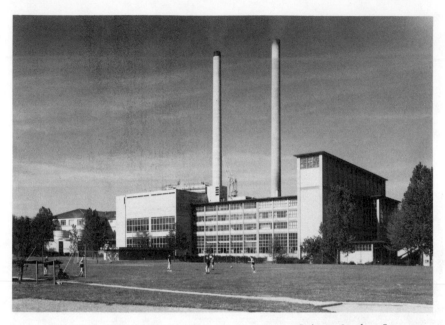

FIG. 34. Resource recovery plant, Berne, Switzerland. In operation since 1954, 200 ton per day capacity, electrical generation, steam/hot water for district heating, industrial processes, and preparation of hospital meals. (Wheelabrator-Frye/Von Roll.)

aesthetic grounds. Therefore, the City of Rotterdam began plans for the new refuse incineration plant.

A government-supported venture, N.V. Afvalverswerking Rijnmond was established in 1968 to operate and maintain the entire municipal resource facility including incineration of solid wastes, distillation of water and electricity generation.

The monies for this project were provided by the Rijnmond Public Authority and the 23 municipalities in the Rijnmond region. Each municipality contributed a share in proportion to its population

and would have all the wastes generated in their jurisdictions transported to the incineration plant in the Botlek. Further financing
of the project was obtained through loans.

Industrial plants in the Rijnmond region can contract with
the N. V. Afvalverswerking Rijnmond for the disposal of their waste
products. Industrial wastes are disposed of at cost to the originator.
Municipal and industrial sources have solid waste disposal priority
over other concerns.

Layout of Facilities

A number of independent components combine to form the
total facility including:

1) The waste reception section: receiving area and storage
pit, scales and a dumping platform for municipal solid wastes and
general industrial and commercial wastes.

2) The irregular waste reception section: receiving area
for liquid and other chemical waste products not readily processed.

3) The waste incineration section: six furnaces each with
a roller-grate combustion process and boiler unit for the incineration
of municipal solid wastes and general industrial and commercial wastes.

4) The irregular waste incineration section: three rotary
furnaces for the combustion of solid and liquid chemical waste; these
units do not include boilers.

5) Electric power generating station: can generate nearly 55,000 kW.

6) Water distillation plant: five evaporators can process almost 600,000 gallons of water per hour.

The plant site was located for easy accessibility by land, water and rail. A canal and receiving area were constructed for handling of wastes arriving by barge.

Urban and Industrial Waste

Municipal solid wastes and certain industrial wastes are incinerated on the mechanical grating system. These are non-hazardous materials and their combustion doesn't result in harm to the 280 employees or to the plant.

These wastes arrive at the facility by way of conventional collection trucks or by barge. The collection trucks discharge their contents from a large dumping platform, 590 ft. long by 98 ft. wide into refuse storage pits. From these bunkers, the wastes are conveyed to five of the six incineration inlet grates by grab cranes and conveyor belts. Alternately, one furnace is always down for maintenance and reserve capacity. Solid wastes arriving via barges are unloaded by grab cranes as the barges moor in the specially constructed harbor.

The Furnaces

Each furnace grate system is comprised of seven rollers, with dimensions of 4 feet in diameter by 13 1/2 feet wide. Combustion air is supplied by a hollow shaft which runs the length of the roller. The

surface of the roller is the grate on which the wastes are combusted
and through which air passes to feed the combustion process. Scrapers,
one in constant contact with the following roller, keep the waste
evenly distributed across the grate and allow for uniform flow of the
burning wastes across the entire grate.

The slag that remains on the final roller is deposited into
a quench tank of water. Drag chains remove the cooled residue and it is
carried by conveyor belt to a slag-treating plant. Here, the slag is
ground and any particles of iron removed. The remaining residue is
sold for use as a road surfacing material, and the ferrous metals are
sold for scrap. Electrostatic precipitators, which are 99 percent
efficient, remove from the flue gases fly ash and other particulates and
the flue gases are then discharged to the atmosphere through two 364 ft.
high stacks.

Chemical Waste Products

The annual capacity of the chemical waste incineration
plant is 70,000 tons, 42,000 tons of liquid wastes and 28,000 tons
of solid wastes. Industries which generate these wastes must comply
to specific restrictions with regards to their disposal. They must
adhere to the incineration plant's timetable, handling criteria, and
amounts to be processed.

Chemical industrial wastes accepted at the plant are those
solid and liquid chemical substances which cannot be incinerated in
a grate furnace. These materials include compounds which require
special handling, such as lead compounds, acrylates, sodium cyanide,

trichloroethylene, insecticide waste, dichloroethane waste, hydrogen
disulfide and phosphorus and arsenic compounds.

Liquid wastes are stored in tanks and solids in containers
before processing. Two of the three rotary furnaces are presently
installed to generate data concerning the process' operation. Each
furnace is an independent unit with two combustion chambers, one for
solid wastes and one for liquid wastes. 1.5 tons per hour of solid
waste are incinerated, and two tons per hour of liquid chemical
wastes are destroyed.

The combustion of these wastes produces gases whose heating
content cannot be utilized by way of transfer to a boiler unit.
Experimentation for the treatment of these gases continues (see below)
as special pollution devices are necessary for their treatment.

A special metering system has been installed for the solid
wastes to ensure proper and uniform combustion. There must be a
constant flow of relatively small quantities of solid waste to the
rotary furnaces, with a maximum charge of 110.23 lbs. A special
accurate metering system was designed with capabilities of delivering
the required charge. The solid wastes are withdrawn from their storage
hopper and conveyed by means of a scraper and conveyor belt to a
gravimetric feed hopper with bottom doors that open automatically.
Material is fed to the hopper until the desired weight is attained.
At this time the furnace conveyor starts up and the hopper's bottom
doors open depositing its contents on the conveyor to be fed into the
furnace. A throughput of 3000 pounds per hour results from this cycle
repeating every two minutes.

The rotary section of the solid chemical waste furnace has dimensions of 13 feet in diameter by 33 feet wide. Water sprayers inject a mist into the furnace to control temperatures to about 1600°F. Higher temperatures would result in the slag beginning to melt and damage to the furnace's refractory lining. Oil can be fired in the furnace should the temperature drop too low. The combustion process in the rotary furnaces is monitored with television cameras and other instrumentation.

The liquid chemical waste is passed to the after-burning chamber of the furnaces from the storage tanks with pressurized nitrogen. Experimentation has been taking place to determine the type of pollution control devices necessary to control the pollutants liberated in the combustion process. Particulate and fly ash emissions are low and control devices for them are unnecessary.

A dry scrubbing method is currently being developed at the University of Technology, the Netherlands to lower pollutant emission rates. Wet scrubbing the gases would reduce flue gas temperatures and velocities such that their dispersal upon exiting the stack would be negligible. A liquid chemical waste pilot plant is under construction where hydrochloric acid and sulfuric acid would be treated prior to combustion.

Stack emissions by the N.V. Afvalverswerking Rijnmond plant are in general controlled to coincide with optimum atmospheric conditions. Wind stability, speed and direction as well as stack parameters are closely monitored to maximize dispersion of the pollutants: 31 monitoring stations are located throughout the plant

area and data is continuously fed to the local air-pollution registration
center which reports to the facility.

The Power Generating Station and Distilled Water Plant

The electric power plant is comprised of three turbine/electric
generators. Each unit has a capacity of 125 tons of steam per hour.
However, under normal operation two turbines receive 250 tons of steam
per hour from the incinerator/steam generator plant to allow one
turbine to be down for maintenance or to absorb peak loads. The combined
electrical output from the three generators is 54,800 kW per hour.

Of the three turbines one is a condensing turbine and two are
back-pressure turbines. After turning the turbines to effect the
generation of electricity, steam leaves the back-pressure turbines
at low pressure (1.5 atmospheres) and low temperature (250°F) at a
rate of 250 tons per hour. The steam then passes to five multistage
evaporators (50 tons per hour each) where salt water is processed,
resulting in distilled water and by-products. The coefficient of
evaporation is 1 to 9, therefore, each evaporator produces 450 tons
of water per hour. Over 76 million gallons of distilled water are
estimated to be produced annually. This water is available to the
local industries and pipelines to supply the water to customers have
already been laid throughout the Botlek region.[32]

SOLID WASTE HEAT RECOVERY IN JAPAN

Japan, like most industrial societies, has had a serious municipal waste disposal problem for several decades, and the problem has been accelerating at a fantastic rate. Over the past five years all of Japan's major cities have shown dramatic increases in the amount of refuse collected; the maximum use seen in Yokohama (200 percent) and the lowest in Tokyo (45 percent).

Although environmentalists were aware of the problems of municipal solid waste disposal early in the sixties, widespread public concern by the Japanese people did not occur until just recently. Furthermore, along with the giant strides in Gross National Product (GNP) over the past decade came giant increases in the amounts of industrial wastes produced.

Tokyo's general practice over the years was to dump the wastes generated by the entire city at a landfill in its Koto Ward. Yet the residents of this ward only recently began to object to the dumping on environmental and hygienic grounds. This opposition

presented a serious obstacle to the future of refuse disposal in the
city, and has resulted in the partial solution of disposing the wastes
generated in each ward within the boundaries of that ward. Alternate
methods always seem to be too costly, unfeasible or environmentally
objectionable.

Public opinion has always been strongly against incineration
and is a hindrance to plant site selection. However, improvements
in the environmental control of incinerator plant effluents, restrictions against refuse generation being initiated in the form
of controls on excessive packaging by department stores and shops,
and recycling of resources, the public is being swayed to consider
incineration as a viable alternative.

Refuse Incineration Plants in Japan

Until the last decade, there was little concern about
the combustion processes of the majority of refuse incineration
plants which were inefficient and usually located in remote areas.
The people involved in refuse incineration were knowledgeable but
their efforts were neglected until the problems almost became over-
whelming. Thus, as public attentiveness grew, so did the number of
incineration plants and the modernization of the older facilities.
Construction of large-scale incineration plants are now underway and
more have been planned.[33]

Table 29 shows the number of incineration plants in each
region of Japan. As of 1970 there were 585 plants of all sizes and
less than half of these utilize the heat of the solid waste com-

TABLE 29 - INCINERATION PLANTS OF JAPAN[33]

Region	Total Incinerating Plants	Plants Utilizing Waste Heat
Hokkaido	6	5
Aomori	12	8
Akita	9	4
Iwate	15	7
Miyagi	7	4
Yamagata	14	7
Fukushima		
Niigata	22	13
Toyama	7	4
Ishikawa	11	4
Fukui	8	2
Nagano	14	9
Gifu	13	4
Shizouka	20	10
Aichi	30	16
Mei	13	5
Ibargi	12	5
Chiba	20	12
Tokyo	29	21
Tochigi	18	10
Gunma	9	8
Saitama	23	12
Iamanashi	7	3
Kanagawa	31	19
Shiga	4	1
Kyoto	9	3
Osaka	30	22
Nara	11	7
Wakayama	5	3
Hyogo	30	7
Okayama	12	7
Hiroshima	12	7
Tottori	5	5
Shimane	7	2
Yamaguchi	12	2
Tokushima	4	7
Kagawa	5	1
Ehim	10	0
Kochi	9	4
Fukuoka	18	9
Saga	8	3
Nagasaki	11	5
Kumamoto	10	5
Oita	7	1
Miyazaki	7	0
Kagoshima	9	3
TOTAL	585	288

bustion process. Furthermore, the waste heat, in most cases, was
utilized for space heating and cooling, and heating water supplies
of the incinerator facility itself or for nearby homes for the aged.

Utilization of the waste heat from incineration has not
been expedited in Japan for a number of reasons, including:

1) The refuse quality (i.e. very low heating values) is
inferior to European wastes and much research has been done on
combustion of the refuse rather than on absorbing the heat from the
process.

2) Unlike European cities, different agencies are
responsible for solid waste disposal and power generation.

3) Marketing the heat product (district heating steam,
electricity) is difficult because industry does not consider it a
viable alternative and/or the remoteness of the incinerator plant
makes utilization of its waste heat impractical.[34]

The need to find alternate economical energy sources as
well as the general advantages to burning refuse on a large scale
which would be necessary for power production, resulted in an
upturn in research in applications to utilize the heat generated
from solid waste combustion. Table 30 shows the solid waste incineration plants with turbo-generators in Japan.

Applications for Refuse Incineration Heat

The Bureau of Sanitation, Yokohama Municipal Office,
Japan has made a number of studies concerning the application of

refuse incineration heat. The numerous applications available
have many limiting factors that must be considered before the
plant reaches the drawing board.

Plant location is exceptionally important. Its proximity
to urban areas is essential where the demand for the waste heat and
the generation of wastes are the greatest. To utilize the waste
heat for district heating a piping network to convey the steam is
necessary, or at least a pipeline to the existing district heating
system. In addition, the incineration plant should be close to the
solid waste collection region to lower transportation costs.

Should plans for the plant include the desalination of
sea-water, a desirable location would be where pollution free
sea-water is available. Environmental impact of the plant must be
considered along with the proper control devices necessary to meet
governmental regulations.

The Yokohama Bureau of Sanitation's Study considered
waste heat applications and the location of incinerator plants to
meet or, rather, to create a demand for the heat product of the
plant. These applications and corresponding plant sites would
include:

1) Desalination/incinerator plants: to be constructed
in pollution free sea-water and regions which can expect water
shortages. (Salt processing plants might then be located near
these plants.)

2) Incinerator/industrial process plants: incinerators
would be built near manufacturing facilities which would use the

TABLE 30 - REFUSE INCINERATION PLANTS WITH TURBO-GENERATORS IN JAPAN

City	Plant	Approved Output (kW)	Type	Evapora-tion (t/h)	Steam Pres. (kg/cm^2 g)	Steam Temp. (°C)	No.	Type	Capacity (kW)
Osaka	Nishiyodo Plant	2700	Corner tube	17	23	350	2	Conden-sation	2700
Tokyo	Setagaya Plant	2500	Corner tube	33	16	203	3	Back pressure	2500
Tokyo	Shakujii Plant	1500	Corner tube	33	16	203	2	Back pressure	1500
Tokyo	Chitose Plant	1700	Forced circu-lation	23.4	12	230	2	Back pressure	1700
Kawasaki	Rinko Plant	1300	Double drum & bent tube	16	16	203	3	Back pressure	1300
Tokyo	Koto Plant	(3000)	Forced circu-lation	27.8	19	240	6	Back pressure	3000
Tokyo	Tamagawa Plant	(2000)	Forced circu-lation	28.2	23	250	2	Back pressure	2000
Tokyo	Ohi Plant	(2500)	Corner tube	27.6	18	209	4	Back pressure	2500
Tokyo	Itabashi Plant	(3200)	Corner tube	28.4	16	203	4	Back pressure	3200
Tokyo	Suginami Plant	(2500)	-	-	-	-	-	-	-
Tokyo	Adachi Plant	(10000)	-	-	-	-	-	-	-
Tokyo	Katsushika Plant	(12000)	-	-	-	-	-	-	-
Tokyo	Tama Plant	Undeter-mined	-	-	-	-	-	-	-
Kawasaki	Tachibana Plant	(2000)	Double-drum & Bent Tube	24.5	16	203	3	Back pressure	2000
Yokohama	Konan	(2800)	Corner tube	35.9	20	211	3	-	2800
Yokohama	Minami-tozuka	(5000)	Double-drum & Bent Tube	65	16	203	3	Back pressure	5000
Yokohama	Undeter-mined	(4500)	-	-	-	-	-	-	-
Kita-kyu-shu	Kogasaki	(3000)	Forced circu-lation	21.5	21	-	-	Back pressure	3000
Sapporo	Shimanopp-oro	(1400)	Forced circu-lation	-	-	-	-	-	1400

As of April, 1973

Steam Pres. (kg/cm² g)	Steam Temp. (°C)	Back Pres. (kg/cm²g)	GENERATOR			Refuse incineration cap. t/d x No. of furnaces
			No.	Capacity (KVA)	Voltage (V)	
23.1	358	0.72	2	2700	6600	200 x 2
10	183.2	0.3	1	3125	6600	300 x 3
10	183.2	0.3	1	1875	6600	300 x 2
10	220	0.5	1	2125	3300	300 x 2
15.5	201.9	0.3	1	1444	6600	200 x 3
17	230	0.5	1	3750	3300	300 x 6
17	230	0.3	1	2500	3300	300 x 2
10	189	0.3	1	3125	3300	300 x 4
12	191	0.3	1	3200	3300	300 x 4
-	-	-	-	-	-	Total 900
-	-	-	-	-	-	Total 1200
			-	-	-	Total 1200
-	-	-	-	-	-	150 x 2
14.5	0.3	1	-	-	6600	200 x 3
12	190.7	0.3	1	2800	6600	300 x 3
14.5	202	0.3	1	5000	6600	500 x 3
-	-	-	-	-	-	400 x 3
18	209	0.5	1	3000	6600	200 x 3
-	-	-	-	-	-	300 x 2

waste heat in their product processing, such as in the manufacture

of ice, concrete, chemicals and even laundry facilities.

        3) District heating and cooling: incinerators would be

constructed in urban commercial districts, suburban residential

districts, and other regions or places where the waste heat could

be utilized, such as heated swimming pools, sport complexes, homes

for the aged, health centers, rural districts.[33]

        The Rinko Plant of the City of Kawasaki is a typical

solid waste incinerator/steam generator and is similar to the

incineration processes previously discussed. Figure 35 is a

steam flow diagram of the plant's electricity generation system.

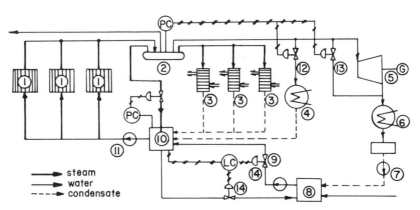

        FIG. 35. Process flow diagram of the steam at the Rinko
plant, Kawasaki.[34] 1, waste heat boiler; 2, steam header;
3, steam air preheater; 4, high pressure condenser; 5, turbo-
generator; 6, low pressure condenser; 7, condensate pump;
8, make-up water tank; 9, make-up water pump; 10, deaerator;
11, boiler feeding pump; 12, main pressure control valve
(primary); 13, ditto (secondary); 14, deaerator level controller.

Table 31 shows two years of operating experience at the plant. Plant refuse intake and power output increased and so did the ratio of power generated to power consumed. Thus the plant is becoming nearly self-sufficient. The number of plant operating days also increased signifying less maintenance problems and lower costs.

The solid waste incineration heat utilization technology is available in Japan. The problem that remains is twofold: 1) to develop a treatment process to effect a solid waste with a high heating value and, 2) to construct incinerator/steam generator large enough to economically utilize the waste heat.

SOLID WASTE HEAT RECOVERY IN CANADA

The need to find alternate solid waste disposal methods and energy sources has encouraged several Canadian metropolitan areas to burn solid wastes to produce power. The factors that have forced Canadian cities to examine incineration/steam generation are the same for urban areas throughout the world: high population density, a scarcity of available disposal sites and restrictive environmental controls. Figure 36 shows the anticipated population growth and its predicted levels of solid waste generation of the greater Quebec area.

With these figures in mind and the fact that Quebec's solid waste disposal problems were becoming acute, feasibility studies and

TABLE 31 - TWO YEARS OF OPERATING EXPERIENCE AT THE RINKO KAWASAKI PLANT[34]

From To	April,1971 March,1972	Monthly Average	April, 1972 March, 1973	Monthly Average
Amount of refuse incinerated (t)	89938	7495	10957	8913
Total amount of power consumed (x 10^3 kWH)	5952	496	6764	564
Amount of power generated (x 10^3 kWH)	4330	361	5683	474
Ratio of power generation de- pendency (%)	72.8	-	84.1	-
Rate of power generated (x 10^3 yen)	16537	1378	21788	1815
Ratio of power generation rate (%)	60.5	-	71.7	-
Number of plant operation days (days)	327	27	338	28
Generator operation hours (hours)	7144	595	7643	637
Operating ratio of power gener- ation (%)	90.9	-	94.3	-
Ratio of single operation (%)	43.0	-	64.7	-

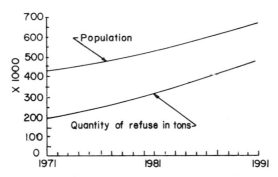

FIG. 36. Predicted population growth and solid waste generation rates.[35]

preparations were made to construct an incineration plant. Many

surrounding communities became involved in the project until

Quebec and 22 surrounding towns formed "The Quebec Urban Community."

Plans were then developed for incineration with heat recovery. The

heat recovery would be in the form of steam and a long term contract

to purchase all the incinerator plant's steam production was signed

with the Anglo Canadian Pulp and Paper Company. The agreement was

contingent upon specified steam parameters, reliability of steam

production and the implementation of a fossil fuel back-up system.

With the sale of steam as a source of revenue, a reduction

in incineration costs was obtained and the capacity of the project

was increased from two to four incinerator units and the plant located

adjacent to the paper mill.

The Plant Process

        Solid waste facility will process 1000 tons per day in
four incinerator units each handling 250 tons of refuse per day.
The refuse has a nominal higher heating value of 6000 BTU/lb and
will be used to produce 81000 pounds per hour of steam at 700 psig
and 700°F from each incinerator unit. Table 32 shows the characteristics of the refuse.

        The system is primarily based on the Montreal incinerator
plant which has been successfully operated for over two years.

TABLE 32 - GREATER QUEBEC AREA REFUSE CHARACTERISTICS[35]

Calorific Value (H.H.V.) BTU/lb		6,000
Composition Percent by Weight:		
Moisture percent		11.45
Dry burnables percent		73.60
Non-burnables percent		14.95
	Total	100.00
Ultimate Analysis:		
Carbon percent		38.67
Hydrogen percent		5.00
Oxygen percent		29.62
Nitrogen percent		0.21
Sulfur percent		0.10
Moisture percent		11.45
Metals and Inerts percent		14.95
	Total	100.00

The grating system for the Quebec plant was fashioned after the
Montreal feed system. The Quebec feed system is a forward-feed
reciprocating grate design. The refuse arrives at the plant via
collection trucks and is dumped into tip and pile combination storage
pit rather than into a deep refuse pit which can cause severe water
pollution problems in the event of a high water table.

Two bucket cranes feed the four feeder hoppers for each
incinerator. Each hopper is equipped with a vibrator to keep the
waste stream moving and a level control device to warn of excess
loading. The refuse passes through the hopper and through the
feeding chute where it is water cooled. Bulky refuse that is too
large to pass through the hopper feed system is first sent through
a size reduction process which is located at one end of the storage
pit.

The grate system is composed of three grates sloped at a
15° angle. The first grate is the shortest grate and used for
refuse drying. The second grate, where most of the combustion takes
place, has mixing arms which spread and tumble the refuse to ensure
complete burning. The third grate conveys the burnt out residue as
well as completing the combustion process. Each grate's speed and
the supply of combustion air can be regulated to effect complete
combustion for the varying heating content of the refuse.

Combustion gases rise within the water cooled furnace
walls and then pass through two evaporators, a superheater,
another evaporator and the economizer of the boiler radiation
section.[35]   Boiler tubes are cleaned by mechanical agitation to

inhibit tubing wear. Figure 37 gives a steam process flow diagram.
Steam is to be delivered at $\pm$ 7% of the paper company requirements,
and a control system has been incorporated to keep the steam within
these limits.

The third grate discharges the burnt out residue into a
quenching tank from which it is tranported to a storage pit by
drag conveyors.   From here it will be trucked to a landfill site
since there are no present plants to implement a ferrous metal
recovery process at the plant. Fly ash and other emissions
recovered from the electrostatic precipitator and tube cleaning

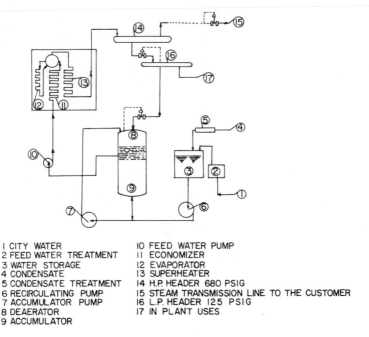

1 CITY WATER	10 FEED WATER PUMP
2 FEED WATER TREATMENT	11 ECONOMIZER
3 WATER STORAGE	12 EVAPORATOR
4 CONDENSATE	13 SUPERHEATER
5 CONDENSATE TREATMENT	14 H.P. HEADER 680 PSIG
6 RECIRCULATING PUMP	15 STEAM TRANSMISSION LINE TO THE CUSTOMER
7 ACCUMULATOR PUMP	16 L.P. HEADER 125 PSIG
8 DEAERATOR	17 IN PLANT USES
9 ACCUMULATOR	

FIG. 37. Steam flow diagram of the Quebec incinerator/steam
generator plant.[35]

process will be conveyed to a storage silo by chain conveyor and
then trucked to landfill sites. Figure 38 shows the entire plant
process flow diagram.

Process waters are to be recirculated in a water sealed
system. Those waters that must be discharged will be treated before
passing to the municipal sewage treatment plant. All steam except
that which is for in-plant use will be transferred to the paper and
pulp company in a steampipe. 50 percent of the steam condensate
will be returned to the incinerator plant for reprocessing. City
make-up water will first be treated and then added to the condensate
to enter the boiler.

Construction costs for incinerator/steam generator, and
projected operating and maintenance costs are given in Table 33.

Although Montreal has an incinerator/steam generator in
operation, the greater Quebec metropolitan area's incinerator/
steam generator is actually the first attempt in Canada of large
scale solid waste heat utilization for industry. The purpose of
the plant is to provide a means of solid waste disposal and to
effectively reduce the costs and pollution contributions by the
generation and sale of steam.

Results from studies of the Montreal and Quebec plants
will be made available for use at the planned facility in Toronto.
An incinerator/steam generator plant is expected to be completed
by early 1977. It will burn about 500 tons of solid waste per year
and generate electricity for one unit of the Toronto utility fed into
the existing power network. The plant process will include solid

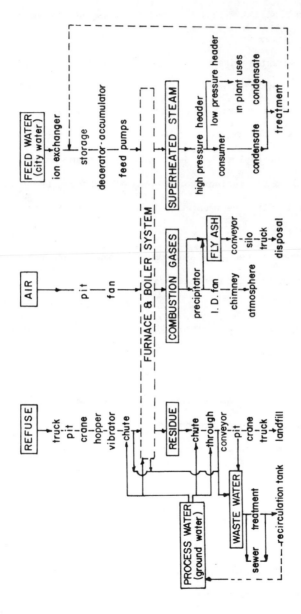

FIG. 38. Quebec Plant process flow diagram.

TABLE 33 - COSTS FOR THE QUEBEC FACILITY[35]

Construction Costs	
Item	Cost
Furnaces, boilers, precipitators, I.D. fans, ash conveyors, ash crane	$11,900,000.
Building, foundations and concrete works	1,247,000
Refuse cranes	426,000.
Pumps and steam turbine	103,000.
Building, steel structure	379,000.
Tanks and heat exchanger	15,000.
Water treatment plant	176,000.
Bunker oil tank	13,000.
Bunker oil heating and pumping unit	16,000.
Control central panel and instrumentation	76,000.
Building, general construction	784,000.
Chimney and fly ash silo	235,000.
Deaerator-accumulator	67,000.
Conveyors system for fly ash	193,000.
Air compressors	56,000.
Power transformers (2)	24,000.
Electronic weighing scale (2)	82,000.
Diesel generator set	33,000.
Emergency steam condenser	253,000.
Temporary services during construction	127,000.
Switch gear panel and bus bars	20,000.
Mechanical works - Steam transmission line	2,175,000.
Access ramps, water and sewer	400,000.
Landscaping and site works	200,000.
TOTAL Construction Cost	$19,000.000.

Operating and Maintenance Costs			
	1975	1976	1980
Tons of Refuse Burned Annually	225,000.	250,000.	325,000.
Amortization	$ 1,768,000.	$ 1,768,000.	$ 1,768,000.
Operating Cost	800,000.	900,000.	1,000,000.
Total Cost/Ton Burned	11.41	10.67	8.51
Revenues from Steam Sale (based on actual price)	1,385,000.	1,458,000.	1,850,000.
Revenue/Ton Burned	6.15	5.84	5.70
Net Cost/Ton Burned	5.26	4.83	2.81

waste separation and metal and glass recovery systems. If the plant proves to be economically and technically feasible, it will be expanded to handle 10 percent of all wastes generated in metropolitan Toronto or 160,000 tons.

CHAPTER 9

CLEAN LIQUID AND GASEOUS FUELS
FROM ORGANIC WASTES

Many methods exist today for the disposal of organic solid wastes, yet each process has its own disadvantages. Furthermore, current environmental regulations limit the use of many methods. A long accepted practice in Europe, but only recently examined in the United States, is the conversion of organic solid wastes into clean gaseous and liquid fuels. There are three factors contributing to the present U.S. attraction to solid waste conversion:

1) Municipal waste disposal problems including incineration pollution effects, a scarcity of landfill sites, and continued increases in solid waste generation.

2) Conventional fuel prices have continued to rise, such that the conversion and use of solid waste fuels has become economically feasible.

3) The conversion of solid waste into gaseous or liquid fuels make it more economically attractive. One large conversion

facility is necessary rather than solid waste processing plants at each application site.

The technology needed to effect solid waste conversion is already available and can be divided into three major processes:

1) Anaerobic digestion - microorganisms oxidize the organic solid wastes producing methane in an oxygen deficient atmosphere, using the oxygen atoms from wastes.

2) Hydrogasification - methane is generated through the reaction of hydrogen with carbon containing compounds.

3) Pyrolytic conversion - gaseous and liquid fuels are the products of thermal decomposition of the organic solid wastes in an oxygen deficient atmosphere.

Other systems have been developed which would recover methane produced from existing landfills through wells sunk deep into the decomposing solid wastes.

Available Feedstock

The U.S. Bureau of Mines estimated that during 1971 the one hundred major metropolitan centers of the United States discarded 71 million tons of organic solids. The gasification of all this waste would have produced over 700 billion cubic feet of methane. The quantities of organic solids in municipal solid wastes are greater than those contained in animal, agricultural and sewage wastes combined. Yet, the available gas from these wastes is relatively small: about 3 percent of the 22.8 trillion cubic feet of

natural gas used by the United States in 1971.

However, the purpose of solid waste conversion is not to

replace conventional fuel, but to supplement it and the entire

energy network. Small percentage energy contributions from a number

of sources can significantly lower the demand for fossil fuels, and

these sources can be many and varied: solid wastes, wind, solar,

geothermal, tides. In addition, solid waste conversion also effects

an alternate waste disposal system, thus benefiting the earth's

environmental and energy cycles.

Anaerobic Digestion

Anaerobic microorganisms digest organics in an oxygen

deficient atmosphere yielding products of carbon dioxide and

methane. A proposed project would utilize this process to provide

1500 ft^3/day of methane gas from 1000 pounds of Milwaukee solid

waste. The project's participants, Allis-Chalmers Inc. of Milwaukee,

Wisconsin, and Waste Management, Inc. of Wak Brook, Illinois, will

construct a pilot plant to acquire data. They believe that the

pilot plant, based on the laboratory work of John T. Pfeffer (University

of Illinois), can produce pipeline quality gas at a cost of $1 per

million BTU. Should this project prove to be economically feasible,

the city of Milwaukee will consider the construction of a full-scale

methane plant to process all its solid wastes.

Figure 39 shows the Pfeffer process. Initially, the

municipal solid waste passes through a shredder and is ground into

particles of uniform size. A magnetic separator then removes the

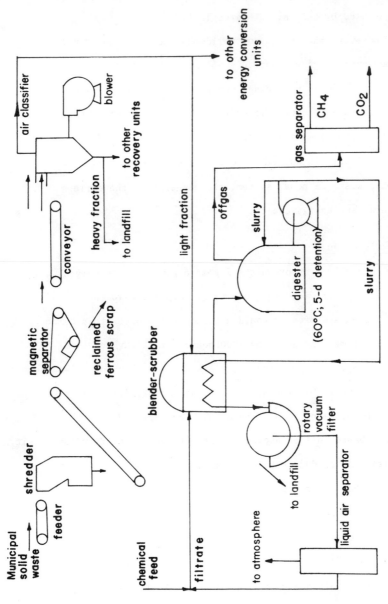

FIG. 39. Pfeffer's solid waste conversion process to produce methane.

ferrous metals to be reclaimed, and the remaining waste is conveyed
to an air classifier which divides the refuse into heavy and light
fractions. The heavier wastes are trucked to landfills or further
classified for resource recovery. The light fraction which is
high in paper content (i.e. cellulose) is mixed with sewage sludge
and/or other chemical nutrients to produce a slurry. The chemical
nutrients or sewage sludge must be introduced since they provide the
required nitrogenous compounds for microorganism growth, and solid
waste usually is deficient in these compounds. This slurry then
has its pH adjusted to 6.7 and temperature raised to 130-140°F and
passed into a digester. These slurry parameters are for optimum
microorganism growth. The slurry circulates in the digester for
5 days. The gas generated by these microorganisms is between 50
and 60 percent methane and the rest carbon dioxide. This gas has a
relatively low heating value, but can be used as is, however, to
obtain a gas of pipeline quality this mixture must pass through a
separator to produce methane and carbon dioxide. Figure 40 shows the
process flow plan of this separator.

The remaining residue in the digester or sludge is a
mixture of lignin, plastics and some unreacted cellulose which, when
it has its water content removed, results in a 75 percent volume reduction for the wastes entering the digester. The properties of this
sludge are being examined for other industrial uses, but should these
studies prove inconclusive the sludge will be dumped as stable landfill.[37,38]

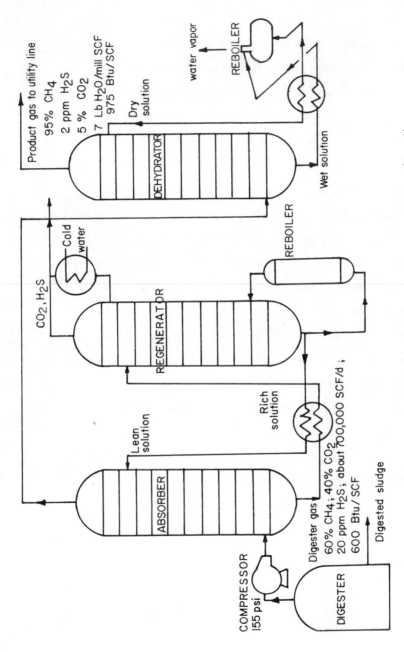

FIG. 40. Methane, carbon dioxide gas mixture separator process flow diagram.

Hydrogasification

Hydrogasification is a process in which hydrogen reacts with organics to generate a methane-rich gas product. This gas is then processed to separate the acid gases and carbon monoxide from the methane. The hydrogasification reaction is highly exothermic, which makes the additional heat unnecessary when solid wastes having high moisture contents are being converted.

Studies have shown that gas generated in this manner had a methane content of about 70 percent and a heating value of over 930 BUT/ft^3. Cattle manure generates a gas product containing nearly 93 percent methane. The cost of the gas is estimated at less than $1 per million BTU.[37]

Research at the Bureau of Mines

Research at the Bureau of Mines on solid waste hydrogasification has proved successful; but since hydrogen is very expensive, alternate methods for solid waste conversion were sought. Researchers at the Bureau of Mines found that a carbon monoxide and water treated cellulose produced a low sulfur fuel.

Cellulose is the major component of organic solid wastes. There are many types of cellulostic wastes including municipal solid wastes, agricultural wastes, sewage sludge, wood, lignin, and animal wastes and they all have been converted to oil. The process involves the high pressure (4000 psig) reaction of cellulose, carbon dioxide and water with temperatures at 350 to 400°C. Various catalysts and solvents are necessary to complete the reaction. Oil yields

obtained have been 40 to 50 percent based on the conversion of over
90 percent cellulose. This oil is a brownish black liquid or a
semisolid at 25°C and has a heating value of 15,000 BTU per pound.
Table 34 shows the sulfur content of oil produced from the conversion
of the various wastes listed.

The processes reactants are as follows:

1) Cellulosic materials; including carbohydrates, wood
wastes (cellulose and lignin), components of municipal solid wastes
(cellulose, carbohydrates, proteins, fats, and other organic materials),
sewage sludge, agricultural and animal wastes along with carbon monoxide
and water can be converted to oil. The plastic content of the municipal
solid waste is not expected to greatly affect the oil processing.

2) Carbon monoxide and water react with cellulose molecules to produce an oil. This process is usually carried out under

TABLE 34 - SULFUR CONTENT OF OIL PRODUCED FROM THE CONVERSION
OF THE LISTED WASTES[39]

Waste	Percent Sulfur
Pine needles and twigs	0.10
Sewage sludge	0.64
Municipal refuse	0.13
Cow manure	0.37
Cellulose	0.003 - 0.2

temperatures of 250 to 400°C to reduce costly carbon monoxide consumption. It also eliminates water and some carbon dioxide and carbon
monoxide to produce a black solid material known as char.

3) A solvent or liquid phase is necessary to prevent the
formation of this solid char. It is also necessary as a vehicle for
the addition of the reactants. Water seems to be the most effective
vehicle, since catalysts readily dissolve in it as do some organic
intermediates of the process and it is very inexpensive. Municipal
solid wastes, sewage sludge, and other substrates probably have
water contents high enough such that no more need be added to the
reaction.

The Conversion Reactions

Cellulose $(C_6H_{10}O_5)$ is the major component of the cell
walls of plants and trees, and therefore of paper and other wood
products. It is made up of long chain glucose units. Starch, like
cellulose, consists of glucose units and is widely distributed in
polysaccharide food reserves stored in the seeds, roots, and fibers
of plants.

The conversion of cellulose to oil is primarily based on
the oxidation of the cellulose molecules and the formation of high
hydrogen-to-carbon ratio molecules. Water and carbon dioxide are
removed from cellulose and other carbohydrates with the addition of
heat. The removal of oxygen to form these high hydrogen-to-carbon
molecules is effected by the addition of carbon monoxide to form
carbon dioxide, by hydrogenation, by various disproportion reactions,

and by combinations of these reactions. As a result of the large
number of reactions possible, the oil produced is comprised of a
complex mixture of different molecules.

Experimental Procedures

0.5 liter and 1.0 liter stainless steel autoclaves were
used to simulate batch operations of the process. White pine wood
chips and newsprint were used as the municipal solid waste source
of cellulose. Carbohydrates were supplied in the form of cellulose,
dextrose and filterpaper. A local dairy farm supplied cow manure
for use in the studies.

The organic solid wastes, catalyst and water were added
to the autoclave at room temperature. Carbon monoxide was added
under pressure and then the autoclave was heated to temperatures of
250-400°C. The temperature of the autoclave was varied to find
the optimum reaction parameters. The heating and cooling of the
system required about 1 hour for temperatures of 250°C, and about
2 hours for temperatures of 350°C and greater.

The products of reaction were removed from the autoclave
with a solvent, and the oil separated from this mixture. Assuming
an average carbon content of the finished oil to be 78 percent, an
oil yield of nearly 57 percent would be effected from the complete
conversion of all cellulose carbon. The major component of the
remaining liquid product is water. However, not all the cellulose
carbon will be transformed into oil as some gaseous products with
carbon content are produced (mainly carbon dioxide).

Figure 41 is a process flow diagram of a continuous operations unit to produce oil from organic solid wastes. System design calls for operation at maximum temperatures of 500°C for flow rates of 100 to 500 g/hr of solid waste slurry and carbon monoxide at a rate of 10 SCF/hr. The carbon monoxide and waste slurry stream is pressurized and heated prior to entry into the bottom of the heated reactor. The liquid and gas products leave the top of the reactor and are separated in a high-pressure recovery process. A back-pressure value continuously releases the gas products while the liquid is collected and discharged into low pressure receivers.

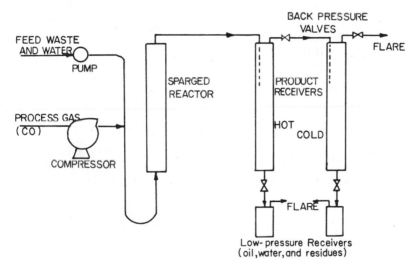

FIG. 41. Process flow diagram of a continuous unit for solid waste conversion to oil.

Pyrolytic Conversion

The pyrolytic conversion process is the third process to
be studied. Pyrolytic conversion, or destructive distillation, causes
the breakdown of materials under a high temperature and in the absence
of air into three components: (1) a hydrogen, carbon monoxide and
methane gas, (2) an oil-like liquid which includes acetic acid,
acetone and methanol, and (3) a nearly pure carbon char. This
method has generated the most interest and technological advances
and has resulted in research by several major corporations. Presently, there are approximately 10-12 different pyrolysis processes under
development.[38,40]

The normal combustion process requires large amounts of air
to complete the burning process and to remove the heat produced. Using
cellulose ($C_6H_{10}O_5$) as the representative organic material the reaction
will take place as follows:

$$C_6H_{10}O_5 \ + \ 6O_2 \ \longrightarrow \ 6CO_2 \ + \ 5H_2O$$

Air supply can be 400 percent above stochiometric requirements. As
a result gas velocities are increased tremendously causing detrimental
effects to air pollution control devices because flow rates through
the device are greatly increased and large particulate loadings are
generated from the air circulating through the burning wastes. These
air pollution control devices make solid waste incineration very costly.

In pyrolysis heat must be added as the volatile compounds are distilled off the solid waste. The pyrolysis reaction is as follows:

$$C + H_2O + Heat \longrightarrow H_2 + CO$$

$$C + CO_2 + Heat \longrightarrow 2CO$$

Carbon reacts with water and carbon dioxide in the presence of heat to produce carbon monoxide. Products of pyrolysis are dependent on time, temperature, pressure, the presence of catalysts, pyrolysis zones, and/or combustion zones within a particular system. However, the principle products of pyrolysis include hydrogen, carbon monoxide, carbon dioxide, methane and other hydrocarbons.[38,41]

The following discussion describes three different pyrolysis processes as developed by three corporations. They are presently under evaluation for full-scale feasibility.

The Garrett Process

Researchers for the city of San Diego, California, have done small-scale experimentation on pyrolysis of solid waste. The research was partially funded by a grant from the Bureau of Solid Waste Management. This research resulted in significant advances in the analysis of the products from pyrolysis of municipal solid waste as well as information on process heat balances. Typical solid waste from San Diego was fed into a pyrolysis reactor. Table 35 shows an analysis of the San Diego solid waste used in this study.

TABLE 35 - ANALYSIS OF SAN DIEGO SOLID WASTE[42]

		Total Material (percent)		
		As Received	Range During Survey	
Combustibles		50.27	55.51 - 43.72	
Incombustibles		49.73	58.28 - 46.33	

		Combustible portion (percent)			
				Dry basis	
	As Re-	Range	Moisture ASTM	With Metal &	Without Metal &
Constituent	ceived	During Survey	D271-58	Glass	Glass
Paper	46.16	50.55 - 40.99	8.23	42.36	50.40
Yard Trimmings	21.14	26.65 - 20.78	51.30	10.30	12.25
Wood	7.48	8.10 - 4.41	10.50	6.69	7.96
Rags	3.46	4.07 - 0.32	7.40	3.20	3.82
Rubber	4.73	4.76 - 3.60	9.74	4.27	5.08
Plastic	0.27	0.52 - 0.25	0.06	0.27	0.32
Garbage	0.81	0.92 - 0.77	57.80	0.34	0.40
Metal	7.64	9.68 - 6.30	0	7.64	0
Glass	8.31	10.38 - 7.67	0	8.31	0
Moisture	-	- - -	-	16.62	19.77
Total	100.00	------------------------------		100.00	100.00

The solid waste is first treated for the removal of glass and metals prior to the start of pyrolysis. The remaining solid waste material was used for analysis in the retort using about 0.4 pounds of refuse per batch. The condensable liquids were passed through a series of cooling traps and condensed. The gaseous products were captured in a gas baloon. The San Diego experiment ran pyrolysis tests at

temperatures ranging from 900° to 1700°F. Gravimetrical analysis
was performed to determine the quantities of char and consensables
and the pyrolysis gases were measured by volume. Table 36 shows the
amount of products produced at the various test temperatures. Qual-
itative analyses were made on the condensables to determine the
organic compounds present and it showed that in addition to water,
the condensable phase included methanol, ethanol, isobutanol,
n-pentanol, tert-pentanol, 1,3-propaneidiol, 1-hexanol, and
ascetic acid. Table 37 shows the results obtained from an analysis
of the char portion and Table 38 presents the results of gas chroma-
tography analysis of the gas phase.

   One of the most significant results obtained from the San
Diego pyrolysis work was the measurement of the heating values of
the products of pyrolysis and evaluation of these in terms of heat
required to sustain the process. Figure 42 is a graph of heat

TABLE 36 - PRODUCTS GENERATED BY PYROLYSIS[42]

Temp. (°F)	Refuse (lb*)	Gases (lb)	Pyroligneous acids and tars (lb+)	Char (lb)	Mass accounted for (lb)
900	100	12.33	61.08	24.71	98.12
1200	100	18.64	18.64	59.18	99.62
1500	100	23.69	59.67	17.24	100.59
1700	100	24.36	58.70	17.67	100.73

* On an as-received basis, except that metals and glass have been removed.

+ This column includes all condensables and the figures cited include 70 to
  80 percent water.

TABLE 37 - PROXIMATE ANALYSIS OF PYROLYSIS CHAR[42]

| | Pyrolyzing temperatures (°F) | | | | |
| | 900 | 1200 | 1500 | 1700 | Pennsylvania Anthracite |
Percent					
Volatile matter (percent)	21.81	15.05	8.13	8.30	7.66
Fixed carbon (percent)	70.48	70.67	79.05	77.23	82.02
Ash (percent)	7.71	14.28	12.82	14.47	10.32
Btu per lb	12,120	12,280	11,540	11,400	13,880

TABLE 38 - GASES GENERATED BY PYROLYSIS[42]

| | Percent by volume at indicated temperatures (°F) | | | |
Constituent	900	1200	1500	1700
H_2	5.56	16.58	28.55	32.48
CH_4	12.43	15.91	17.73	10.45
CO	33.50	30.49	34.12	35.25
CO_2	44.77	31.78	20.59	18.31
C_2H_4	0.45	2.18	2.24	2.43
C_2H_6	3.03	3.06	0.77	1.07
Accountability	99.74	100.00	100.00	99.99

recovery and utilization from pyrolysis products. To obtain a self-
sustaining pyrolysis process and a thermal efficiency of 50 percent,
the pyrolysis temperature should be 900°F. The char content could be
adjusted to provide additional heat to the process. It should be
noted however that these results are based on batch processes and that
a continuous pyrolysis process is expected to have much better self-
sustaining characteristics.

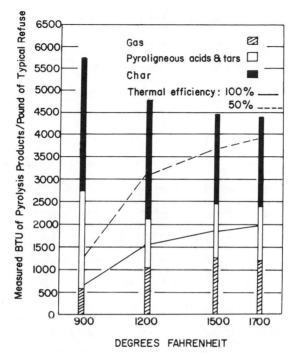

FIG. 42.   Recovery and utilization of solid waste energy
through pyrolysis.[42]

Process Description

The Garrett process is a low-temperature flash pyrolysis.
It converts ground up organic material into a high viscosity and high
oxygen content fuel oil, combustible char and gas. Initially the
municipal solid waste is shredded into rather coarse 3-inch particles
requiring a shredder output of between 50-60 Hp-hours/ton. From the
primary shredder the refuse is transported to an air classifier and
separated into a light fraction consisting of paper and plastics, and
a heavy fraction consisting of glass, metals, wood and rock. This

separation procedure reduces the inorganics in the light fraction to less than 4 percent.

The light solid waste fraction passes through a drying and screening process to reduce the inorganic content further and then is shredded to a very fine particle size (less than 24 mesh), requiring a shredder output of between 50-60 Hp-hours/ton. These fine organic solid waste particles enter the pyrolytic reactor and flash pyrolysis takes place at temperatures of around 900°F.

The heavy fraction passes through a magnetic separator for removal of ferrous metals and a mixed-color glass cullet is removed by froth flotation in a 99.7% purity. The remaining char is trucked and dumped at landfill sites. The heat required for pyrolysis is derived from combustion of the fuel gas and a fraction of the char produced. The Garrett system can produce over 1 barrel of oil/ton at 4.8 million BTU/BBL or 6,000 SCF of gas at 800 BTU/ft^3 without requiring supplemental fuel sources to supply energy for the process. Figure 43 is a block diagram of the Garrett process flow.[25,41,42,43,44]

Envirochem Landgard System

As previously described the Landgard system was developed by Monsanto Research Corp. through the operation of a one ton per day pilot plant in Dayton, Ohio and a 35 ton per day plant in St. Louis county from the fall of 1969 through the fall of 1971. Results from these operations enabled Envirochem to (1) demonstrate the continuous pyrolysis of typical unclassified municipal solid waste, (2) acquire solid waste handling experience, (3) gather the

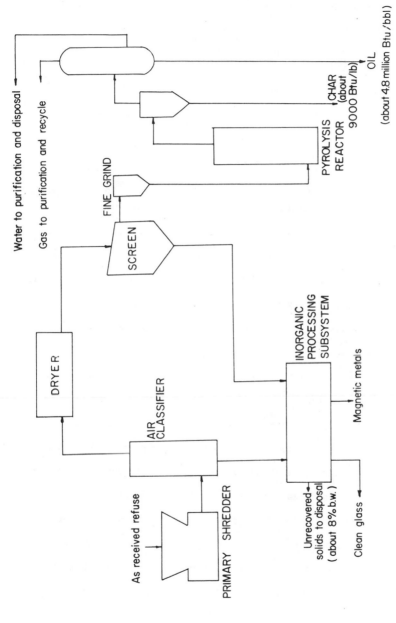

FIG. 43. Garrett pyrolysis process.

data necessary to design and construct large scale facilities, and
(4) develop a solid waste residue resource recovery process.

Process Description

The Envirochem Landgard process is the low-temperature
pyrolysis of the municipal solid waste's organic compounds through
partial oxidation with air in a rotary kiln to produce char and
combustible gases. Unclassified municipal solid waste is milled
to produce a uniform particle size and conveyed to a bin. From
the storage bin the shredded waste is fed continuously into the
rotary kiln. Shredded waste enters one end and fuel oil enters the
opposite end of the kiln. The flow of gases and solids is countercurrent to expose the feed to higher temperatures as it passes through
the kiln. Thus, first drying and then pyrolysis occurs. The final
residue is contacted with temperatures of 1800°F just before release
from the kiln. In order to maximize pyrolysis action the kiln is
designed to expose solid particles to a uniform temperature gradient.

The hot residue released from the kiln enters a water-
filled quench tank and passes into a flotation separator. The
material that floats off is a carbon char slurry. It is thickened
and filtered to remove the water, and conveyed to a storage area
before being trucked to landfill sites. The putresible content of
this residual char is about 0.1 percent which makes it ideal for
landfill operations. The heavy non-buoyant material is conveyed
along the bottom of the flotation separator to a magnetic separator
for removal of ferrous metals which are deposited in a storage area

for railcar or truck shipment to scrap dealers. The balance of the heavy material, or glassy aggregate, passes through a screening process and then stored on-site for use in road construction.

Pyrolysis gases are drawn from the kiln to be piped to local utilities or industries for use as low BTU value gas to be combusted with oil, natural gas, or coal. Table 39 gives an

TABLE 39 - LANDGARD SYSTEMS GASEOUS FUEL ANALYSIS[45]

Nitrogen	69.3
Carbon Dioxide	11.4
Carbon Monoxide	6.6
Hydrogen	6.6
Methane	2.8
Ethylene	1.7
Oxygen	1.6
Solids Content	1-2 GR/DSCF
Moisture Content	19% By Volume
Heat Available	250 $\bar{M}$ BTU/HR (HHV)
(From 1000 TPD)	50 $\bar{M}$ BTU/HR (SENSIBLE)
	300 $\bar{M}$ BTU/HR

analysis of gaseous fuel obtained from this process. Figure **44**
is a block diagram of the Landgard system gaseous fuel option.[41,45,46]

Union Carbide Oxygen Refuse Converter (Purox[TM] System)[7,9]

Background

The Purox system converts unclassified municipal solid
waste into a clean burning fuel gas and an inert residue. It uses
oxygen as a supplement to the pyrolysis reaction of the combustibles
of the refuse to effect an environmentally acceptable fuel gas. It
further provides high temperatures needed to form molten metal and
slag from the noncombustible fraction of the solid waste. This
resultant residue is inert and only 2 percent of the volume of the
originally charged refuse.

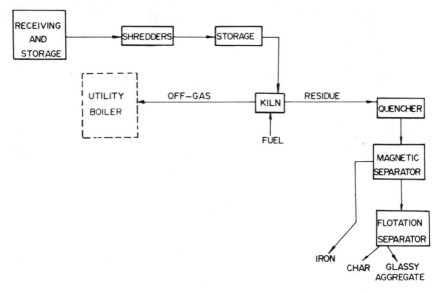

FIG. 44. Landgard systems gaseous fuel option.[45]

Process Description

The Purox process combines the rather unique advantages of
pyrolysis (the generation of useful and valuable by-products) and
high temperatures (the melting and fusing of the metals and glass).
These are the effects of using oxygen rather than air in the conversion step.

Figure 45 shows the vertical shaft furnace on which the
process is dependent. Solid wastes enter the top of the furnace
and oxygen is injected into the bottom. Char forms from the refuse
and oxygen reacts with it generating high temperatures in the order
of 2600° to 3000°F. These temperatures are needed in the hearth
to melt and fuse the metals and glass. This molten metal and glass
sludge gravimetrically drains continuously into a water quench tank
where it hardens into a granular material.

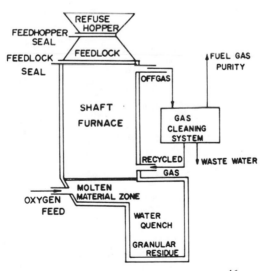

FIG. 45. Vertical shaft furnace.[46]

The reaction of oxygen and char generates hot gases which rise up counter to the flow of the falling refuse and pyrolyzes the refuse as it cools. Near the upper portion of the furnace, the gases and refuse intermingle causing the gases to cool further and the refuse to be dried. The gases then exit the furnace at about 200°F. These gases consist of large amounts of water vapor, oil mist, and small quantities of other constituents. The undesirable components are removed in the gas cleaning system.

The gas that leaves this cleaning process is a clean burning fuel comparable to natural gas with a heating value of about 300 BTU/ cubic foot. The fuel characteristics are shown in Table 40. Its sulfur and nitrogen compounds are virtually non-existent as it can be

TABLE 40 - PUROX PYROLYSIS FUEL GAS COMPOSITION[46]

Constituent	Volume %
CO	49
H_2	29
CO_2	15
CH_4	4
$C_2H_2^+$	1
N_2 + Argon	2
Total	100

used directly as a fuel supplement to any fuel consuming operation.
The emissions generated by combustion of this fuel are within air
pollution regulations. Fly ash emissions have been measured at
0.008 grain/cubic foot or an order of magnitude below federal
regulations. It can be combusted without the addition of costly
modifications to the boiler system.

        The Union Carbide process results in a net energy output.
The final gas generated represents 83 percent of the fuel value
of the unclassified municipal wastes entering the conversion system.
A minimum amount of this fuel gas is used by the process to generate
steam to heat the plant buildings, and to heat and maintain the
auxiliary combustion chamber at operating temperatures. Thus, the
remaining fuel gas or approximately 75 percent of the fuel energy
of the refuse is available for outside combustion sources. The
usage of available energy from a 1000 ton per day refuse processing
facility is shown in Table 41.

        The residue that is left from the noncombustible fraction
of the refuse is sterile and compact because it has been melted and
fused to eliminate any biologically active material and to a minimum
volume. **Sanitary** landfill techniques are unnecessary for its
disposal. It is suitable for construction fill material. Depending
on the noncombustible content of the incoming municipal solid waste
a final volume reduction of 97 to 98 percent can be effected. For
the most efficient incinerators a volume reduction of 90 percent or
less is considered good design. A process flow diagram for the Purox
system is given in Figure 46.

TABLE 41 - USAGE OF AVAILABLE ENERGY AT A 1000 TON/DAY
OXYGEN-REFUSE CONVERTER FACILITY[46]

	BTU/Hour	Percent
Available Energy in Refuse[a]	416,000,000	100
Energy Losses in Conversion Process[b]	70,000,000	17
Energy Available in Fuel Gas	346,000,000	83
Fuel Gas Uses:		
Process Steam	16,000,000	4
Building Heating	10,000,000	2
Energy to Maintain Auxiliary Combustion Chamber at Operating Temperature	7,000,000	2
Net Energy Available in Fuel Gas	313,000,000	75
Electric Power Generation	30,000 KW(c)	
Electric Power Used In Plant	5,000 KW	
Electric Power Available for Export	25,000 KW	

(a)   Based on a refuse heating value of 5000 BTU/lb., this is calculated as
      (5000) (2000) (1000)/24.

(b)   Includes latent heat of moisture in refuse, sensible heat of fuel gas,
      heat content of molten slag and metal, and heat leak.

(c)   Based on combustion of the net fuel gas in a gas utility boiler with
      35% thermal efficiency.

Fuel Gas

        The fuel gas generated by the Purox process can be used as

a supplementary fuel in existing boilers or other fuel consuming

operations. Table 42 compares the fuel gas of solid waste origin

to some common hydrocarbons and carbon. The fuel gas has a lower

heating value than methane (natural gas) or propane or butane.

The problem arising from this relatively low heating value is the

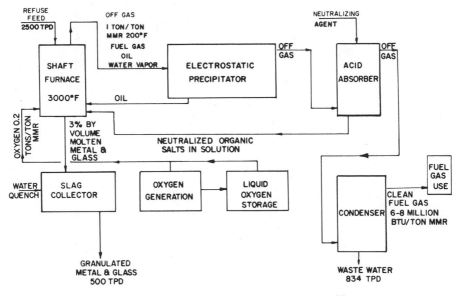

FIG. 46. Purox process flow diagram.[47]

necessity of increased compression power for this fuel gas relative
to methane. However, this is a minor economic factor because the
pressures involved are relatively low.

The combustion air requirements for the fuel gas per
unit of heat release are less than the methane or the other fuel air
requirements. Table 43 shows the combustion air requirements for
the fuel gas. Therefore, the capacity of existing air blowers should
be sufficient for the combustion of fuel gas without modifying the
boiler.

The quantity of combustion products generated per unit of
heat release is also lower for the fuel gas than for methane or the
other fuels. This is due directly to the lower combustion air re-

TABLE 42 - A COMPARISON OF SOME COMMON FUELS AND THE PYROLYSIS GENERATED FUEL GAS

	Fuel Gas	Methane	Propane	n-Butane	Carbon
Lower Heating Value (BTU/SCF) (1)	286	910	2,312	3,010	
Compression Power (KWH/MM BTU) (2)	5.7	1.8	0.6	0.5	
Combustion Air Requirements (SCF/MM BTU) (3)	8,300	10,500	10,300	10,300	10,400
Volume of Combustion Products (SCF/MM BTU)	10,500	11,600	11,200	11,100	10,400
Heat Release/Volume of Combustion Products (BTU/SCF)	95	86	89	90	96

(1) Standard cubic foot dry, as measured at 60°F and 1 atmosphere pressure. Heating value calculated at 18°C.

(2) Gas compressed to 35 psig from 1 atm., 100°F with 75% efficiency.

(3) Based on the air needed to convert the fuel to CO_2 and H_2O - no excess air.

TABLE 43 - UNION CARBIDE (PUROX) OXYGEN REFUSE CONVERTER COMBUSTION CHARACTERISTICS OF FUEL GAS

Basis: One Standard Cubic Foot of Fuel Gas (Dry)

Fuel Gas Composition		Oxygen Requirement	Volume of Combustion Products, SCF		
Constituent	Volume %	Per SCF of Constituent Fuel Gas	H_2O	CO_2	N_2 & Ar
CO	49	0.5	0.245		0.49	0.921
H_2	29	0.5	0.145	0.29		0.545
CO_2	15				0.15	
CH_4	4	2	0.080	0.08	0.04	0.301
C_2H_2	1	3	0.030	0.02	0.02	0.113
N_2 & Ar	2	-				0.020
Totals	100		0.500	0.39	0.70	1.900

Volume of Combustion Air Required

Per SCF of Fuel Gas = 0.500/0.21 = 2.38 SCF
Per Thousand BTU = (1000) (2.38)/286 (LHV) = 8.34 SCF

Volume of Combustion Products

Per SCF of Fuel Gas = 0.39 + 0.70 + 1.900 + 0.03* = 3.02 SCF
* Moisture Content of Fuel Gas, Assuming Saturation @ 35 Psig, 100°F

Per Thousand BTU = (1000) (3.02)/286 (LHF) = 10.5 SCF

quirements for the fuel gas as seen in Table 43. Existing induced
draft fans and air pollution control devices should not be adversely
affected by the combustion of this fuel. This is important at plants
that burn dirty fuels and sized pollution control equipment must
remain effective.

Finally, the Purox process fuel gas combustion products
release more heat per unit volume than the other fuels listed in
Table 42, signifying that fuel gas flame temperature and heat transfer
characteristics are similar to the other fuels.

In general the fuel gas of solid waste origin will be a
small fraction of the total fuel requirements of a typical utility
boiler. Thus, variations in production fuel gas flow and heating
values can be absorbed through adjustment in the base fuel flow rate.
This would require a relatively simple control system.[25,41,46]

Landfill Gases

Another potential source of clean gaseous fuels from
organic solid waste are the gases generated at sanitary landfills.
NRG NuFuel Co., a subsidiary of NRG Technology, Inc., has developed
a method of withdrawing and purifying landfill gases. A 12-month
study involving the drilling of three test-wells at different areas
of the Los Angeles County Palos Verdes Landfill site has begun.
Should the tests show gas is available in quantities and qualities
(1,000 ft^3/min, 50% methane) of economically feasible proportions,
NRG will install a molecular sieve purification station on the site

to collect and then sell the gases to Southern California Gas Co.
Further landfill testing is being discussed for California, Arizona,
Illinois, New Jersey, New York, and Pennsylvania. A landfill is
economically feasible if it can produce 3.5 - 7 lb. of methane per
pound of landfill and there is a nearby market for the gas.[37,47]

CHAPTER 10

PLASTICS RECYCLING: AN ENERGY SOURCE

Plastics have such a widespread usage in our society today that they are taken for granted. The market for plastics has increased rapidly over the past three decades. They are used quite readily in the construction of many items, both as substitutes for more costly materials and as durable and adaptable replacements.

Marketing practices by the plastics industry have resulted in the development of numerous industrial, commercial and domestic applications. Plastics can be formed into any shape, made in any size, dyed to any color, produced in any thickness, density and strength and still be inert to many substances. As a result, they are ideal for containers and other packaging, toys, tools, utensils, construction materials, and coatings and laminants.

Origin and Occurrence of Plastics

Petroleum sources are the origin of most plastics. Close to 5 percent of all the petroleum extracted from the earth each year is allocated to the petrochemical industry and about 1 percent of

this 5 percent goes into generating plastics. Production of plastics
begins at the oil refinery and natural gas purification plant where
the crude oil and gases are separated into their many components.
Petrochemical plants receive these processes' products and produce
the basic entities of plastic-ethylene and propylene. These basics
are then processed to effect the final product. Ethylene is polymerized into polyethylene. Ethylene and chlorine are united to produce
vinyl chloride which is then polymerized to polyvinylchloride (PVC).
Ethylene and benzene are processed into styrene monomer and then
polystyrene. Propylene is polymerized into polypropylene. Most
of the plastics produced are merely the basics being built upon
through various processes.

The quantities of plastics in municipal solid waste are
growing rapidly. Plastic content of a typical waste is presently
about 2 to 3 percent, yet not too long ago it was only 1 to 2 percent. This is directly proportional to plastic consumption which is
also growing rapidly. Plastic production from 1960 to 1970 increased
by an annual rate of 11.8 percent.[48]

Although current petroleum prices have caused shortages and
increased plastic prices, plastics consumption is accelerating since
plastics are still more economical than other materials and new applications are always being developed. This is seen in the new automobiles,
in the beverage packaging industry which now plastic coats its bottles
to inhibit breakage and reduce glass requirements of the bottles.
Table 44 shows the trends over the past decade of the generation of

TABLE 44   PACKAGING WASTE GROWTH OVER THE PAST DECADE
1965-1975[49]

Category	Thousands Of Tons/Yr 1965	1975	Percent Increase	% Of 1965 Total	% Of 1975 Total	Change 1965- 1975 (%)	% Change Of Total
Glass	8,060	11,840	32	18.01	19.18	+1.17	+ 6.5
Foils	113	185	64	0.25	0.30	+0.05	+20.0
Wood	5,872	4,057	-31	13.12	6.56	-6.56	-50.0
Metal	6,583	8,641	31	14.71	14.00	-0.71	- 4.8
Wax	318	450	42	0.71	0.73	+0.02	+ 2.8
Fiber and Paper	22,444	34,063	52	50.15	55.18	+5.03	+10.0
Plastics	1,364	2,490	82	3.05	4.03	+0.98	+32.2
Total	44,754	61,726		100.00	100.00		
	(38% Growth)						

packaging wastes. The estimate for 1975 shows a plastic waste increase
of 82 percent over 1969. At these rates the plastic fraction of the
total municipal solid waste will probably reach 6 percent sometime in
the 1980's and continue to rise. Note Table 45 for estimated production tonnage of the various plastic types for the next decade.

Main Types of Plastics

        The term "plastics" seems to signify those materials that
are flexible and unbreakable. However, there are many rigid and
brittle plastics and these physical characteristics are based on
the particular plastic's chemical characteristics. Generally all
plastics are long chain molecules with an almost infinite number of
repeating entities and these are known as polymers.

TABLE 45 - PRODUCTION OF THE VARIOUS PLASTIC TYPES BASED
ON AVERAGE ANNUAL GROWTH RATES OF 10% and 12%[50]

	Annual Growth %	Production in Tons 1975	1980
Average Annual Growth Rate of 10 Percent			
Polyolefins	11	745,000	1,240,000
PVC	11	535,000	900,000
Polystyrenes	10	250,000	400,000
ABS	10	29,000	45,000
Other Thermoplastics	7	165,000	230,000
		1,724,000	2,815,000
Thermosets	7	560,000	800,000
Grand Total		2,284,000	3,615,000
Average Annual Growth Rate of 12 Percent			
Polyolefins	12.66	900,000	1,600,000
PVC	12.0	575,000	1,000,000
Polystyrenes	11.5	300,000	530,000
ABS	13.5	40,000	95,000
Other Thermoplastics	11.33	210,000	425,000
		2,025,000	3,650,000
Thermosets	10	675,000	1,100,000
Grand Total		2,700,000	4,750,000

There are two basic categories for plastics: 1) thermoplastic materials which can be remelted, reformed and reused, and
2) thermosetting materials which cannot be reformed.

Thermosetting plastics are comprised of a cross-linked
network of molecular chains which are bonded to one another at the
cross. These cross-link bonds reinforce and strengthen the thermosetting materials, making them more heat and corrosion resistant and
enabling them to retain their shape until decomposition temperatures
are reached.

Thermosetting plastics comprise nearly 20 percent of all
plastic production including the various phenolics, polyesters, and
epoxies.

The remaining and most common plastics are the thermoplastic
type. Table 46 shows an analysis of the Western World's use of the
basic plastic types of the thermoplastic materials available--the most
widely used today are as follows:

TABLE 46 - AN ANALYSIS OF PLASTIC TYPE USAGE IN THE WESTERN WORLD[51]

Polyolefins	29%
Polyvinyl Chloride	23%
Polystyrene	13%
All Others	35%

1) Polyethylene (PE), high and low density, is probably
the most common plastic used. It is opaque, is usually resistant
to varying degrees of impact and is available in varying degrees of
flexibility. Polyethylene is an economical manufacturing material.
A major portion of the plastic bottles and films are made with polyethylne because it is inert to many chemical corrosives and climatological effects. Various sources estimate PE to be 29 to 38 percent
of total plastic waste.

2) Polypropylene (PP) has chemical properties very similar
to polyethylne. Its primary advantage over polyethylene is its very
good resilience properties. Polypropylene packaging is steadily increasing.

3) Polyvinyl chloride (PVC) is easily recognized in
packaging by its clarity and flexibility. In the packaging industry
PVC is a natural substitute for glass. However, it is not limited
to this application as it is widely used in shoes, piping, sheeting
and guttering, fabrics, etc. The common product name vinyl is actually
an abbreviation for PVC. Various sources estimate PVC to be 20 to 32
percent of total plastic waste. Table 47 gives an analysis of PVC end
uses.

4) Polystyrene (PS) is different from the other plastics
because the products it products are more rigid. It is relatively
inexpensive, it is transparent, and brittle, but has numerous applications as plastic containers, toys, appliance components, etc. Polystyrene is estimated to be between 13 to 21 percent of the total plastic
waste.[48]

TABLE 47 - TYPICAL ANALYSIS OF PVC END USES[50]

Cables and Wires	10%
Thin Film and Artificial Fabrics	17%
Flooring and Tiles	13%
Other Foils	11%
Rigid Rods and Shapes	26%
Laminates and Other Rigids	9%
Rigid Sheets	12%
Bottles (Rigid)	2%

Plastic coatings and adhesives for packaging materials have increased quite rapidly. For the 1965 to 1975 decade we see from Table 48 an increased usage in all types of plastics but a nearly 62 percent rise in polyethylene and related product production. This trend is further noted in the steady rise in plastic consumption in the flexible plastic packaging field (see Table 49) as we observe a 56 percent rise in polyethylene and copolymer consumption, a 250 percent rise in polypropylene consumption, and a 360 percent rise in PVC consumption. Table 50 shows rigid plastic consumption.

Increased plastic wastes present many problems with disposal as we shall now discuss. However, increased plastic content of a municipal solid waste increases the heating value of each pound of refuse. With the appropriate amounts of plastics this heating value may reach 8000 Btu/lb.

TABLE 48 - PLASTICS AS COATINGS AND ADHESIVES FOR
PACKAGING MATERIALS[49]

	1965**	1975**
Polyethylene and Copolymers (ethylene/vinyl acetate and ethylene/ethyl acrylate)	162	263
PVC (can linings, cap linings)	12	13
PVC (paper and board)	2	3
PVC (glass bottle coatings)	3	4
Polyvinyl Alcohol	10	13
Polyvinyl Acetate	41	55
Styrene-Butadiene	3	4
Acrylic	3	4
Saran	10	20
Other*	20	25
Total	266	404
Wax	318	450

*Nylon, polycarbonate, fluorocarbon, ionomer, polyurethane, epoxy, cellulose nitrate, special compounds.

**Thousand Tons

Present Methods of Disposal

The problems of plastic disposal have become acute and will worsen in the future. The manufacture of plastic disposable items continues to increase the amounts of plastic waste. Presently, the

TABLE 49 - PLASTICS FOR FLEXIBLE PACKAGING[49]

Plastic Film	1966**	1970**	1975**
Polyethylene and Copolymers	425	600	750
Polypropylene	30	50	75
Cellophane	200	175	150
Polyvinyl Chloride	35	85	125
Polyvinylidene Chloride	10	7	5
Linear Polyester	4	5	5
Polystyrene	5	8	8
Cellulose Acetate	2	2	2
Pliofilm	8	8	8
Other*	5	10	15
Total	724	950	1,143

*Nylon, polycarbonate, fluorocarbon, ionomer, polyurethane.

**Thousand Tons

Note: Growth will be in bags -- multiwall and liners, overwraps, shrink overwraps, and film/tray shipping containers. There will be increased production of specialty film laminates, and coated structures, i.e., nylon/polyethylene/saran.

most common and technically advanced method of solid waste disposal are landfills, composting, and incineration. However, each disposal method presents numerous problems associated with the safe and efficient disposal of plastics.

TABLE 50 - PLASTICS FOR RIGID PACKAGING[49]

	1965	1975
Low Density Polyethylene	10	20
High Density Polyethylene	160	475
Polystyrene (including impact)	155	250
Polyvinyl Chloride	4	50
Polypropylene	10	68
ABS	5	30
Urea	12	10
Phenolics	7	5
Cellulosics	9	10
Other*	2	25
Total	374	943

*Nylon, polycarbonate, phenoxy, sulfones, acetals, acrylates, urethane cushioning.

Notes: Includes -- thermoformed containers from sheet, rigid plastic boxes, containers, and lids, blister and rigid skin packs, closures, foamed cases and inner packing. Excludes -- beverage cases, plastic bottles, and tubes.

Sanitary Landfill

        Sanitary landfill has been regarded throughout history as the most acceptable method for solid waste disposal including plastic wastes. Yet, there are many disadvantages to this method when plastics must be landfilled.

Covering discarded plastics with soil or other fill simply removes their visibility since plastics do not decompose. Certain plastics are resistant to chemical corrosion and biological attack. There seems to be no species of microorganisms that can biologically degrade plastics at a sufficient rate to promote effective disposal. Thus, when buried in a landfill plastics can remain intact for many years or for our time perspective, infinitely. Dumping plastics at sea results in the same effects as landfills -- no degradation. Furthermore, compaction of solid wastes prior to landfill is essential and plastics pose the problem of the relative incompressibility. As a result of their inherent flexibility and resistance to permanent compression and deformation many difficulties are experienced in the attempt at volume reduction of some polymers, and with the scarcity of landfill sites this loss of landfill space cannot be afforded.

Composting of Plastics

The utilization of solid wastes as useful soil conditioners has on occasion proved to be an attractive approach to the solid waste disposal problem. However, aside from the technical problems associated with composting, composting techniques applied to plastics are ineffective because these materials are largely biologically inert.

Incineration of Plastics

Incineration of solid wastes is widely accepted because it provides for a 80 to 90 percent volume reduction, and it eventually

utilizes the waste heat generated. There are many unresolved problems
in the design and operation of conventional incinerators with plastics.

In a complete and proper combustion process polyethylene,
polypropylene and polystyrene and large quantities of excess air will
yield products of carbon dioxide and water. The combustion of polyvinyl chloride generates fumes of hydrogen chloride (HCl), a toxic gas,
which must be controlled. HCl in combination with moisture forms HCl
acid and is highly corrosive. However, HCl is generated not only from
the burning of PVC but also from burning normal municipal wastes such
as paper, salt in food, grass, wood, leather, etc. Performance of
conventional incinerators has not been entirely satisfactory because
in practice combustion of plastics is often incomplete, producing chain
fragments as well as many other problems.

Effective and complete combustion in an incinerator depends
upon the maintenance of proper temperature, time, oxygen flow, flame
turbulence, and fuel distribution. Low temperatures, below 1400°F,
may result in excessive smoke emissions by burning organic waste.
Significant amounts of unoxidized hydrocarbons are found among in-
completely combusted products of plastics. Furthermore, the emission
of large quantities of carbon monoxide pose potential air pollution
hazards. Grate systems can be rendered ineffective when the heavier
thermoplastic polymers melt into a sticky molten mass on the grate and
block the air supply and inhibit refuse flow.

Probably the most important problem with conventional
incinerators in regards to plastics is the high calorific value of the
plastics, which is approximately 15,000 to 16,000 BTU's per pound.

Most municipal solid waste incinerators are designed primarily to burn materials that release much less heat then plastics between 5000 and 10000 Btu/lb such as cellulosic materials (paper and wood) and the low calorific constituents of municipal refuse. In general, all that is necessary to properly combust these materials is oxygen and heat. However, plastics, with their chemical differences, high molecular weights, and hydrocarbon natures, sometimes inhibit the efficient combustion of materials in a conventional incinerator. The combustion of plastics produces relatively high temperatures, above 2400°F, which are damaging to conventional refractory furnace linings and gratings. Also, these high temperatures can promote reactions that lead to the formation of pollutant emissions such as hydrogen cyanide.

Due to the non-homogeneous nature of plastics in the solid waste stream when fed into the incinerator furnace, local hot spots are formed causing slagging and spalling damage to the refractory linings. Slagging is the building up of ash deposits on the refractory bricks and spalling is the chipping or breaking away of the brick surface caused by temperature shocks, the rapid heating and cooling of the bricks. The large amounts of heat generated by the combustion of plastics can also effect damage to the grate system and many air pollution hazards.

Present levels of plastic wastes found in municipal solid wastes (2 to 3.8 percent of total) do not result in unusual difficulties; however, solid waste managers are becoming concerned because it is believed that an increased plastic content to 3 or 4 percent will begin

to cause incineration problems, and concern is mounting over the growing quantities of plastic waste and possible future problems.

Control of the heat transfer reactions in a conventional
municipal solid waste incinerator is as such impractical, but many of
the problems inherent in conventional incinerator plastic combustion
process can be eliminated through increased processing of the raw
solid wastes, i.e. shredding and mixing, and improved incinerator
design to handle increased plastic loads.

Materials Recovery of Plastics

Essentially, the recovery of plastics for recycle material
from mixed solid waste does not presently take place. There are a
great many economical and technical problems accompanying the salvaging,
sorting, and separating of plastic materials from solid wastes. Iden-
tification of the immense number of different plastic formulations
(there are over 700 different grades of polyethylene alone) is extremely difficult and segregation by resin type is practically impossible,
and probably unfeasible, preventing their reuse after they are discarded.
It has been noted that short chain polymers might be recoverable through
thermosetting the plastic by pyrolysis.

The recovery of plastic materials from mixed solid waste for
recycle is further complicated by the industry demand for high quality
scrap. The economics for such a process do not warrant the underwriting
of a recycling project. Manufacturers prefer more economical raw materials which are readily available.

The fundamental problem which arises from recycling plastics is
their synthetic origin. Unlike metals' processing, which begins with

impure ore that is refined and purified, plastics' processing begins
with high purity polymers that are developed with various nonresinous
additives, dyes, and reinforcements additives. There is generally no
technology for reprocessing plastics, and a practical means of effecting
contiminant removal is still largely non-existent.

Yet another problem in recovering plastic waste is the pre-
dominance of laminants which contain both thermoplastic and thermosetting materials. The molecular structure of these materials is
incompatible, making fabrication of products from these mixed wastes
impossible.

Energy Recovery From Plastics

The rapid growth of plastic usage and wastes, and the
technical and economic problems associated with separating it from
raw municipal solid wastes, suggest that plastic wastes may be best
utilized through the recovery of their latent energy value as an
energy source. The utilization of energy from waste plastics also
effectively eliminates the plastic waste disposal problem. Plastic
available in mixed municipal wastes has a higher heating value of
nearly 16,000 BTU/pound, which is the highest heating value of any
component in municipal wastes. The average heating value of municipal
solid wastes is around 5500 BTU/pound. Thus, this high heating value
makes plastic extremely useful and important for energy recovery. Since
the proportion of plastics as well as paper in solid wastes continues
to rise, the available energy to be recovered is also rising. The
methods available for energy recovery from plastic wastes include:

1) heat recovery of incineration processes

2) refuse derived fuel (RDF) for use in power plant furnaces

3) pyrolysis

Heat recovery through incineration is similar to the solid waste combustion processes previously discussed and is subject to the same corrosion and environmental problems, but nevertheless, is better than no recovery at all.

Refuse derived fuel (i.e. Ecofuel) currently is being considered by many to be the solution to municipal solid waste disposal and resource recovery. The municipal solid waste is processed and the light combustible fraction of the refuse is recovered and refined into a viable fuel source which can be combusted in large industrial and utility boilers either as the primary fuel or a supplemental fuel, usually resulting in substantial cost savings. Refuse derived fuel has a higher heating value (HHV) of 7000 BTU/pound compared to coal whose HHV is about 10,000 BTU/pound. Its processing entails the removal of noncombustibles through several stages of classification, control of the moisture content, and multi-stage fine shredding of the combustible fraction. Available data has not determined specific effects of plastic waste on refuse derived fuels, but from a full heating standpoint they should be beneficial because as the plastic content rises so does the heating value of the refuse derived fuel.

As a result of the high degree of treatment the municipal solid waste is subjected to effect refuse derived fuel; the derived fuel is a homogeneous and thoroughly mixed compound. Localized hot spot problems are not encountered in the burning of this fuel compound, thus eliminating many of the plastic inherent problems of conventional incinerators.

Polyvinyl chloride presents the only potential limiting factor of effectively recovering energy from plastics through the refuse derived fuel method. Even though the PVC content is but a small proportion of the total solid waste volume (approximately 0.5 percent), PVC presents corrosion and pollution problems that will continue and grow as the PVC content of the solid waste significantly increases in the future.

Pyrolyting systems, currently in the large pilot plant stage, continue to be a very promising solid waste disposal and energy recovery method, and can become the ultimate solution for the future. The burning of the organic portion of mixed municipal waste in an oxygen deficient atmosphere results in the chemical breakdown of the long chained polymers to lower molecular weight compounds.

As the organic analysis of the municipal solid waste varies, the products of pyrolysis varies. This output is also dependent on waste flow rate, temperature, temperature gradient with the pyrolytic reactor, yet the nature of the influent controls the products of destructive distillation. CO, CO_2 and O_2 will evolve from oxygenated materials. N_2 or ammonia will evolve from nitrogenated materials such as proteins and polyurethanes. Methane and ethane production will increase as the plastic content of the waste increases.

The recycling of plastics can be effected through pyrolysis. Raw materials, or a gas or liquid fuel can be produced by pyrolytic action. Plastic wastes that can be separated from municipal solid

wastes or plastic wastes generated as scrap are pyrolyzed into a
product with a chemical composition similar to crude oil. Table 51
shows this comparison with respect to individual components. This
plastic pyrolyzation product can be used as a raw material for the
petrochemical industry or further refined to be utilized as a fuel.

TABLE 51 - COMPARISON OF AN ACTUAL CRUDE OIL TO A PYROLYTIC
EFFECTED SYNTHETIC CRUDE OIL[52]

	Crude Petroleum	Plastic Pyrolyzate
Saturated Hydrocarbon	Yes	Yes
Straight Chain	Yes	Yes
Branched Chain	Yes	Yes
Cyclic	Yes	Possibly Traces
Unsaturated Hydrocarbons	Yes	Yes
Straight Chain	Traces	High
Branches Chain	No	Yes
Aromatic	Yes	Yes
Alkylbenzene	Yes	Styrene Toluene
Styrene Telomers	No	Yes
Naphthalenes	Yes	No
Phenanthrenes	Yes	No
Sulfur Compounds	Yes	No

The pyrolyzation of polyvinyl chloride entails the use
of neutralizers which react with HCl and render it harmless. Sodium
and calcium carbonate, as well as tin compounds, are used as additives
to the pyrolytic process to neutralize the HCl.

Therefore, the processing of a synthetic crude oil is technically feasible. The building of a plant for production is dependent on
(1) the availability of a separation method to classify plastics and
other wastes, and (2) economics comparable to natural crude oil.
(A factor that should be considered when figuring economics of the
energy recovery from wastes is that solid wastes are being disposed
and this cost should adjust the basic economics.)

However, the recycling of plastics and production of synthetic
crude oil requires that a plastic waste not contaminated with other
municipal waste components be used as the pyrolytic reactor feed stock.
The reuse of certain plastics is highly selective, for the resultant
products have high purity requirements. Thus, in many instances, these
processes can be impractical.

Pyrolysis is best suited for energy recovery and disposal of
municipal solid wastes and other wastes with large plastic contents.
Many factors contribute to this thesis including:

1) Pyrolyis destroys the wastes at high temperatures leaving
a biologically inert residue.

2) The process generates gases containing high contents
of alkaline compounds, especially ammonia, which neutralize the acid
products generated by certain wastes. This includes the neutralization
of hydrogen chloride which evolves from the degradation of PVC.

3) Pyrolysis produces a gas, similar to coal gas, of consistent quality that can readily be utilized. Plastics contribute significantly to this gas quality.

4) The process is available not only to solids but also to liquids or to solids that melt at relatively low temperatures, i.e. plastics. The molten plastic does not cause detrimental effects to the pyrolysis mechanics.

5) The process can be run at full scale without major problems.[50]

CHAPTER 11

WASTE OIL AVAILABILITY
AND TREATMENT PROCESSES

Waste oil is any oil which has been processed and no
longer exhibits the necessary chemical and/or physical properties
required of a virgin oil. There are two primary sources of waste
oil: 1) automobile and truck waste oils, and 2) aviation and
industrial waste oils. Other sources of waste oils include oil
spills, tank cleaning, ballast water, and a variety of other
petrochemical and chemical activities. The automotive waste oils
are primarily of crankcase drainings which constitue over 90 percent of this type of waste oils. The remaining 10 percent consists
of transmission fluids, gear lubricants, hydraulic oils from automotive and construction equipment and relatively small quantities
of kerosene and other solvents used in equipment servicing. Industrial and aviation waste oils are comprised of a variety of oils
and oil emulsions. These are wastes generated by metal working,
lubrication of industrial equipment, hydraulic and circulating
systems, railroad diesel engines, turbine lubrication, and facilities that overhaul aircraft engines. The remaining waste oils are

oils which have been used in transformers and heat transfer and re-
frigeration equipment.[53]

In a highly industrialized society large quantities of
waste oils are generated annually. The quantity of waste oil generated is directly proportional to oil consumption. The U.S. Bureau
of Mines estimated the lubricant demand for 1972 to be 2.2 billion
gallons. Table 52 shows a sales estimate for the various lubricating oils. The Environmental Protection Agency[54] has estimated that
over 50 percent of lubricating oil consumption or about 1.25 billion
gallons of waste oils are generated annually. Automotive lubricating
oils account for nearly 50 percent of the total lubricating oil sales
and also represent the greatest potential environmental impact through
improper collection techniques, heavy metal content, and reprocessing/
re-refining requirements. This chapter therefore will be concerned
primarily with this waste oil source.

Available Waste Oils

According to the American Petroleum Institute (API),[55] of
the 1.1 billion gallons of crankcase oil consumed in the United States,
approximately 68 percent is removed from automotive combustion engines
as waste oil. The remaining oil is lost either through leakages or
combustion during engine operation. Thus, there are about 800 million
gallons of waste automotive oils available annually. Other national[56]
and regional[57,58,59,60] estimates have this volume varying by as much
as 50 percent (i.e. 400 to 730 million gallons) depending on the
source of estimation such as motor vehicle statistics, service station
statistics, and automotive lube oil market statistics. The Environ-

TABLE 52 - LUBRICATING OIL SALES (Millions of Gallons)[54]

	Sales
Automotive Lube Oils	
Service Stations	270
Garages, Auto Supply Stores	60
New Car Dealers	102
Retail Sales for Commercial Engines	90
Auto Fleet and Other Lube Oil Uses	136
Factory Fills (Auto and Farm Equipment)	60
Discount Stores	168
Commercial Engine Fleets	200
	1086
Industrial and Aviation Lube Oils	
Hydraulic and Circulating System Oils	325
Metal Working Oils	150
Railroad Engine Oils	60
Gas Engine Oils	62
Aviation and Other	137
	734
Other Industrial Oils	
Process Oils	310
Electrical Oils	57
Refrigeration Oils	10
	377
Lube Oils Purchased by U.S. Government	37
Grand Total	2234

mental Protection Agency has estimated the quantity of waste oils to
be 616 million gallons,[61] based on many factors some of which are
included in Table 53. EPA estimates of the remaining waste oils generated are given in Table 54.

TABLE 53 - ESTIMATE OF FACTORS FOR CONVERTING AUTOMOTIVE
SALES INTO WASTE OIL QUANTITIES[61]

Service Stations

    70% of Oil Sold is Used for Changes.
    Oil Drained is 90% of Filled Capacity.
    70% x 90% = 63% of Oil Sold - Waste Oil Generated.

Garages and Auto Supply Stores

    Assume Average is Same as Service Stations (63%).

New Car Dealers

    100% of Oil Sold is Used for Changes.
    Oil Drained is 90% of Filled Capacity.
    100% x 90% = 90% of Oil Sold = Waste Oil Generated.

Retail Sales for Commercial Engines

    Assume same as Service Stations (63%)

Automotive Fleet and Other Lube Oil Uses

    Assume 50%, allowing for two-cycle engines and internal use, e.g. fuel,
    by commercial and governmental fleets.

Factory Fills, Automotive and Farm

    Assume 90% recovery as in automotive service centers.

Oil Bought at Discount Stores

    Assume Same as Service Stations (63%).
    Assume 35% of waste oil generated finds its way to service stations.
    63% x 35% = 22% of Oil Sold = Waste oil generated at service stations.

Numerous studies have been conducted to determine the
quantity of waste crankcase oil collected for recycle. These
studies and two specifically, one survey of Maryland and the other
of Pittsburgh, indicate that a discrepancy between what is actually
available and what is estimated to be available. The problem with

TABLE 54 - SOURCES AND QUANTITIES OF WASTE OIL[61]

	Millions Of Gal/Yr
Automotive Lubricating Oil	616
Industrial and Aviation Lubricating Oil	394
Other Industrial Oils	87
U.S. Government Lubricating Oils	18
Marine Oil Transporation Losses and Spills	209
Other losses in Production, Refining, Transportation and Use	1156
Total	2480

these surveys and the estimates they effect is that they are dependent on rather ambiguous factors such as the reckonings of a
service station owner.   Nevertheless, all information available shows
that at least 50 percent of the crankcase oil that is estimated to be
available for recycle, is available.   Therefore, 200 to 400 million
gallons of waste crankcase oil is collected annually.   However,
Table 55 shows the operating throughput of selected waste oil re-
refiners, and although not all waste oil reprocessors are identified,
it does note that only 100 million gallons of automotive waste oils
are actually re-refined.   The remaining waste oil is utilized as a
fuel, road oil, or discharged to the environment as shown in Table 56.
Large quantities of waste oil are burned along with No. 4 or 6 oil as
a supplemental fuel for industrial furnaces.   Utilization of waste oil

TABLE 55 - SURVEY OF SELECTED OIL RE-REFINERS: PROCESSES AND CAPACITY[61]

| Company | Processes | | | Capacity (gpd) | | Waste Crankcase Oil* |
	Dehydration	Acid-Clay	Other	Design	Operating	Feed gpd
No. 1	X	X	-	-	5,000	4,000
No. 2	X	X	-	2,200	2,100	2,000
No. 3	X	X	-	23,000	12,000	12,000
No. 4	X	X	Batch Basis	-	9,000	-
No. 5	-	-		-	-	-
No. 6	X	X	3/4 Capacity 1/4 Capacity	20,000	20,000	16,700
No. 7	-	-	Hi Pres.Dist.	130,000	65,000	-
No. 8	X	X	-	62,500	50,000	-
No. 9	X	X	W/Steam Strip	4,300	3,200	2,750
No. 10	X	X	-	6,500	3,250	3,000
No. 11	-	-	Dry,HiSpdCent	12,500	12,500	11,200
No. 12	X	X	-	5,250	5,250	4,700
No. 13	-	-	Caustic Clay	17,000	17,000	-
No. 14	X	X	-	20,000	10,000	-
No. 15	X	X	Steam Strip	-	8,300	4,100
No. 16	X	X	-	12,500	10,500	5,000
No. 17	X	X	-	-	20,000	12,000
No. 18	X	X	Steam Strip	12,500	12,500	-
No. 19	X	X	-	7,500	7,500	-
No. 20	X	X	-	8,300	8,300	6,250
No. 21	-	-	Caustic Clay	40,000	30,000	-
No. 22	X	X	-	33,000	20,000	-
No. 23	X	X	-	17,000	10,000	7,000
No. 24	-	-	Caustic Clay HC1 Act.Clay	25,000	5,000	4,500
No. 25	-	X	-	6,700	2,000	2,000
No. 26	X	X	-	-	1,000	1,000
No. 27	X	X	-	3,600	2,000	2,000
No. 28	X	-	-	40,000	20,000	-
No. 29	-	-	-	33,000	-	-
TOTAL				636,850	372,400	100,200
AVERAGE				28,000	14,000	

*Other sources not listed.

TABLE 56 - WASTE OIL FINAL END POINTS[61]

	Millions Of Gal/Yr.
Directly to the Environment	978
Processing Wastes to the Environment	72
Road Oil, Dust Control, Runoff, Asphalt Leaching, Land Disposal, Residue to the Environment	281
Sub-Total-To The Environment	1,331
Directly For Use As Fuels	358
Use for Processing Plant Fuels	49
Fuels Sold by Oil Processors	621
Sub-Total-Fuels	1,028
Road Oil and Dust Control-Rapid Biodegradation	38
Re-refined Oil Sales	83
Sub-Total-Other	121
Total for Petroleum	2,480

as a supplemental fuel conserves conventional fuels but results in

higher atmospheric lead emissions. Combustion of all the waste crank-

case oil would effect the emission of from 7000 to 15,000 tons of

lead to the atmosphere annually if just 50 percent of the lead contained

in the waste oil found its way into the flue gas. These emission rates

are but a fraction (4 to 8%) of the lead emissions from automobile

exhausts, which are presently being phased out. The combustion

of unprocessed waste oil not only pollutes but also represents a
loss of a natural resource. The lead content of waste oils ranges
up to one percent concentration and of waste oil distillation residues in concentrations up to 10 percent. Thus, the concentrations
of lead found in waste oils is equal to or greater than concentrations
of lead found in many ores mined commercially.

     Roads are treated with waste oils to inhibit dust circula-
tion, particularly in farming areas, and significant quantities of
waste oils are used in season. Waste oil is less costly than a
special oil compound, and thus is favored for its short term economics.
However, the waste oil is easily weathered from the road and does not
have a lasting effect on the formation of dust. Also, due to runoff
leaching it is a potential water pollution problem.[62] Therefore, for
environmental and resource reasons, other methods of dust control must
be developed.

     A fraction of the waste oil is utilized in hot road asphalt,
either in its raw waste state or as a distillate cut from the waste oil.
This is apparently an acceptable application for waste crankcase oil
providing asphalt specifications permit.

     Before we discuss the processing and applications of waste
crankcase oil, we will briefly describe certain waste oil characteristics and note their relation to lubricating oil specifications.
The environmental effects of waste oil mismanagement will also be
discussed.

Waste Oil Characteristics

        Waste oil is a term which encompasses a rather hetero-
geneous group of oil lubricants, including waste crankcase oil,
transmission fluids, differential gear lubricants, hydraulic oils,
and small quantities of other solvents. Waste crankcase oils
usually constitute the major portion of waste oil lubricants. Depending on their ultimate usage these waste crankcase oils can contain the following substances: (1) a moderate amount of sulfur
which is inherently present in lubricating oils, (2) various
functional additives to improve the original product properties
including: viscosity index improvers, detergents, detergent in-
hibitors, dispersants and antioxidants, (3) iron "fines" effected
from engine fretting and wear, (4) gasoline components, oxidized
materials, atmospheric dust and other products of combustion such as
water, (5) sedimentary materials derived from internal engine
deposits, and (6) water and other contaminants introduced into waste
oil storage tanks.[53]

        A study conducted by the Petroleum Analytical Research Corp-
oration[63] resulted in typical characteristics and composition of waste
oil to be as shown in Table 57. In addition to the metals noted in
Table 57, numerous other trace metals are present in waste oils. A
typical analysis of waste oil trace metal constituents is given in
Table 58. From this analysis it is evident that appreciable amounts
of lead, barium, calcium and zinc are found in most used motor oils.

TABLE   57 - TYPICAL AUTOMOTIVE OIL WASTE COMPOSITION[63]

Variable	Value
Gravity, °API	24.6
Viscosity @ 100°F	53.3 Centistokes
Viscosity @ 210°F	9.18 Centistokes
Flash Point	215°F (C.O.C. Flash)
Water (By Distillation)	4.4 Volume %
BS & W	0.6 Volume %
Sulfur	-.34 Weight %
Ash, Sulfated	1.81 Weight %
Lead	1.11 Weight %
Calcium	0.17 Weight %
Zinc	0.08 Weight %
Phosphorous	0.09 Weight %
Barium	568 ppm*
Iron	356 ppm*
Vanadium	5 ppm*

* ppm = parts per million

The actual trace metal content is sometimes dependent on the analytical procedures used on the waste oil sample, and large differences
in values have been obtained by emission spectroscopy and atomic
absorption. The metal content of the individual samples and the

TABLE 58 - ANALYSIS OF WASTE OIL SHOWING TRACE METAL CONSTITUENTS[63]

	Parts Per Million			
Metal	Average		Range	
Aluminum	16	10	-	30
Copper	28	5	-	120
Iron	361	150	-	800
Lead	1524	960	-	2070
Silicon	46	10	-	240
Tin	30		30	
Sodium	64	20	-	110
Barium	269	10	-	900
Calcium	1772	1160	-	2690
Zinc	1111	560	-	1610
Magnesium	155	10	-	420

many types of waste oils can vary widely. According to these data lead concentrations may vary by an order of magnitude. The uncertainties of these analyses have important implications for waste oil processing, since in order to treat the oils for various metals contents, a processor must know the throughput oils' composition. Figure 47 compares the properties of waste oil and virgin fuels. Tables 59 and 60 list the SAE viscosity number specifications and DOD Mil Specs on typical virgin lubricating oils.

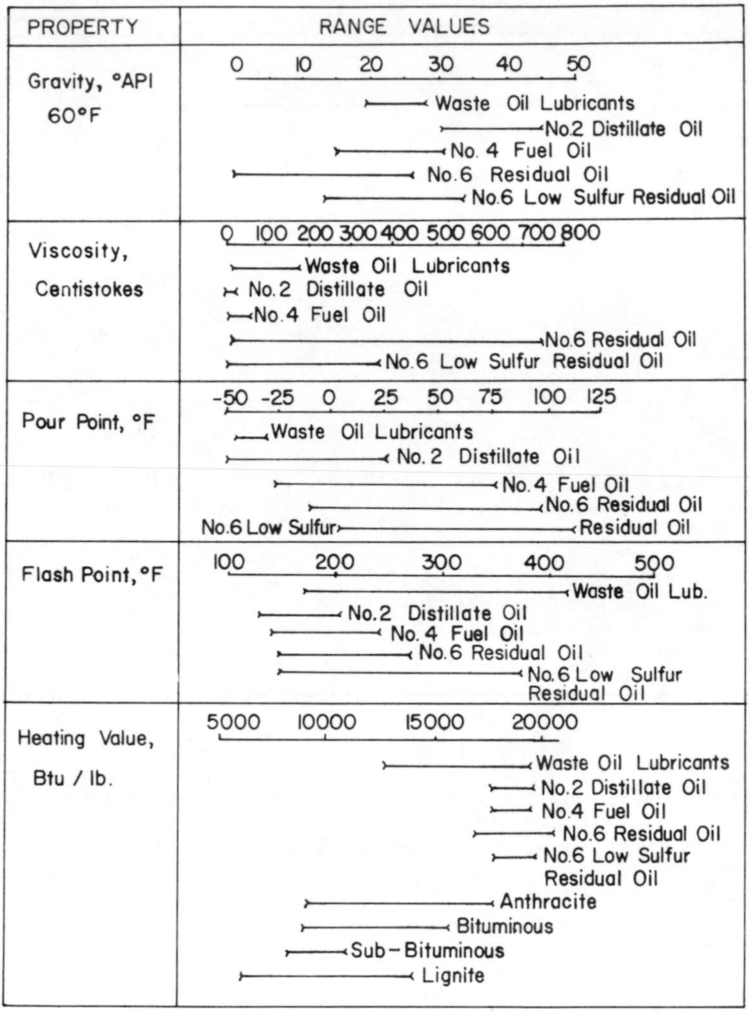

FIG. 47. Comparison of waste oil and virgin fuel property ranges.

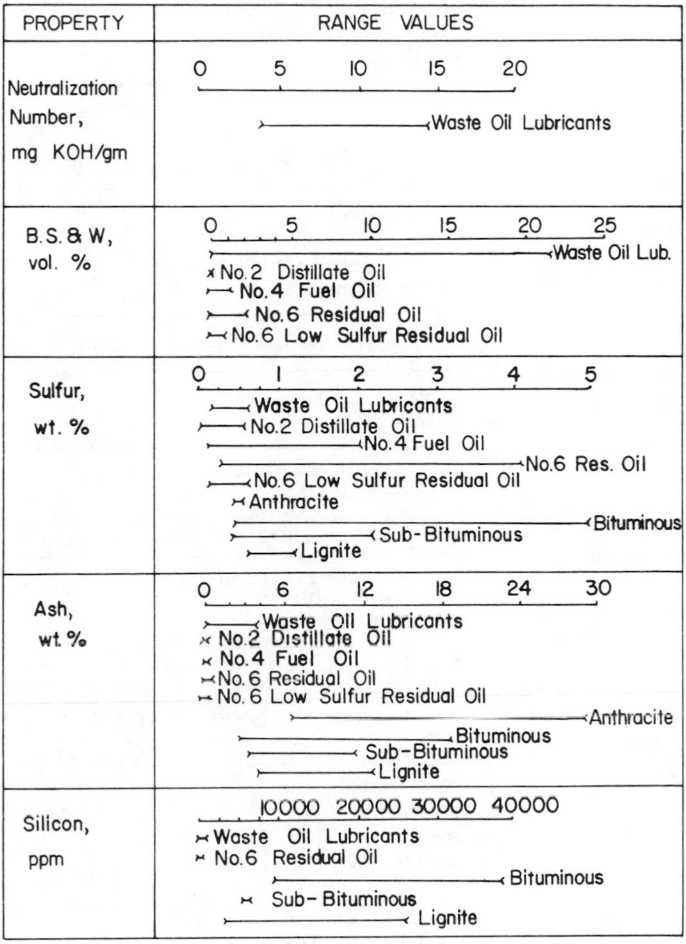

PROPERTY	RANGE VALUES
Neutralization Number, mg KOH/gm	0 5 10 15 20 ›————————————‹ Waste Oil Lubricants
B.S. & W, vol. %	0 5 10 15 20 25 ›————————————————————‹ Waste Oil Lub. × No.2 Distillate Oil ►—‹ No.4 Fuel Oil ›——‹ No.6 Residual Oil ►—‹ No.6 Low Sulfur Residual Oil
Sulfur, wt. %	0 1 2 3 4 5 ›——‹ Waste Oil Lubricants ›——‹ No.2 Distillate Oil ›———————‹ No.4 Fuel Oil ›————————————————————‹ No.6 Res. Oil ›——‹ No.6 Low Sulfur Residual Oil ►‹ Anthracite ›———————————‹ Bituminous ›—————‹ Sub-Bituminous ›——‹ Lignite
Ash, wt. %	0 6 12 18 24 30 ›——‹ Waste Oil Lubricants × No.2 Distillate Oil × No.4 Fuel Oil ►‹ No.6 Residual Oil ►‹ No.6 Low Sulfur Residual Oil ›————————————————————‹ Anthracite ›——————————‹ Bituminous ›—————————‹ Sub-Bituminous ›————————‹ Lignite
Silicon, ppm	10,000 20,000 30,000 40,000 ►‹ Waste Oil Lubricants × No.6 Residual Oil ›————————————————‹ Bituminous ►‹ Sub-Bituminous ›——————————‹ Lignite

FIG. 47. (continued)

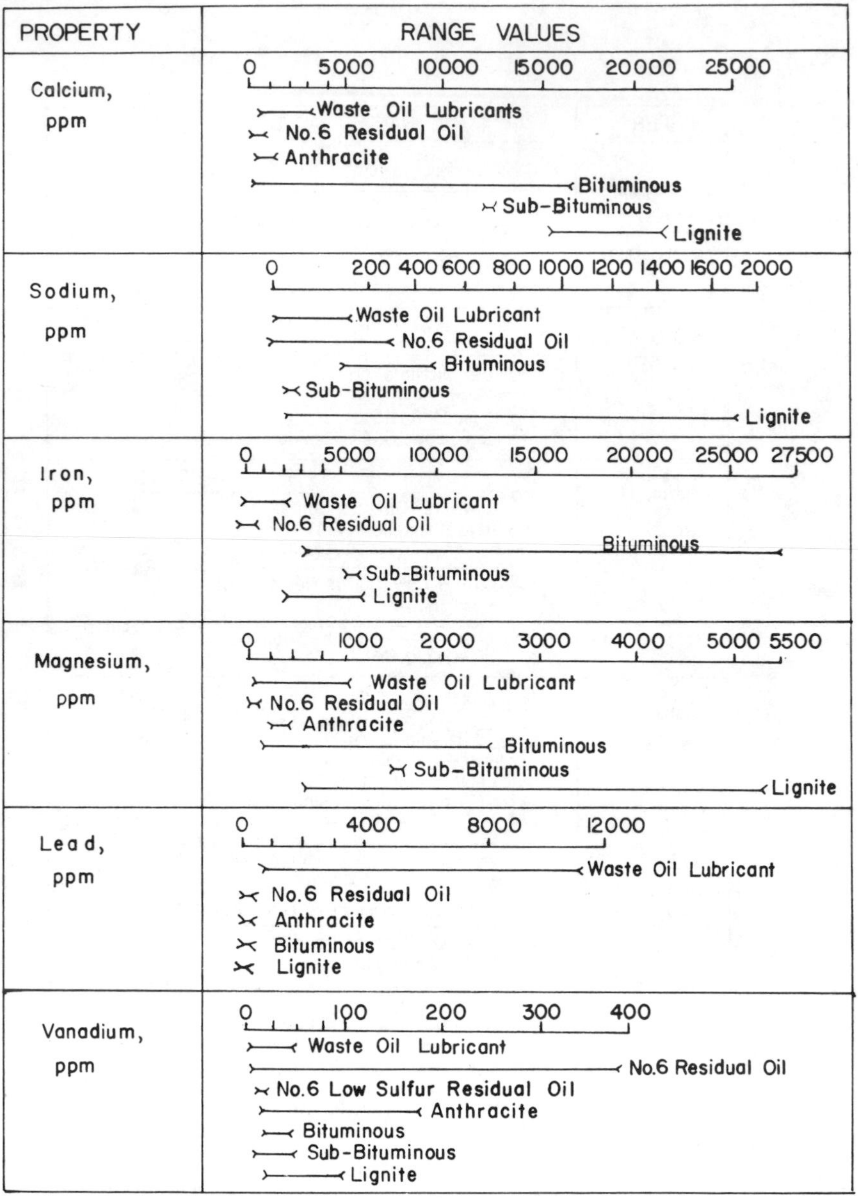

PROPERTY	RANGE VALUES
Calcium, ppm	
Sodium, ppm	
Iron, ppm	
Magnesium, ppm	
Lead, ppm	
Vanadium, ppm	

FIG. 47.   (continued)

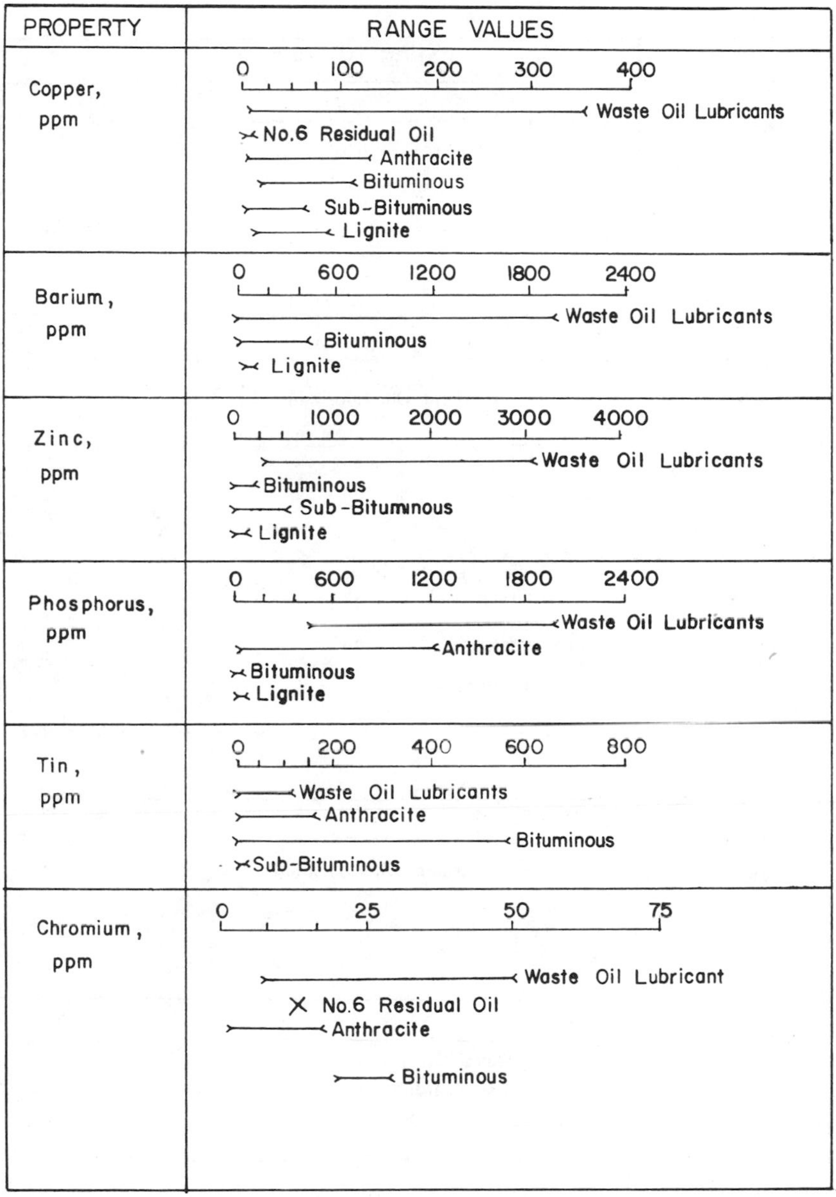

PROPERTY	RANGE VALUES
Copper, ppm	0 100 200 300 400 ⊢—————————————< Waste Oil Lubricants >—< No.6 Residual Oil >——————< Anthracite >————< Bituminous >——< Sub-Bituminous >——< Lignite
Barium, ppm	0 600 1200 1800 2400 >——————————————————< Waste Oil Lubricants >————< Bituminous >—< Lignite
Zinc, ppm	0 1000 2000 3000 4000 >—————————————< Waste Oil Lubricants >—< Bituminous >———< Sub-Bituminous >—< Lignite
Phosphorus, ppm	0 600 1200 1800 2400 >——————————————< Waste Oil Lubricants >——————————< Anthracite >—< Bituminous >—< Lignite
Tin, ppm	0 200 400 600 800 >———< Waste Oil Lubricants >——————< Anthracite >———————————< Bituminous >—< Sub-Bituminous
Chromium, ppm	0 25 50 75 >——————————< Waste Oil Lubricant X No.6 Residual Oil >——————< Anthracite >——————< Bituminous

FIG. 47.   (continued)

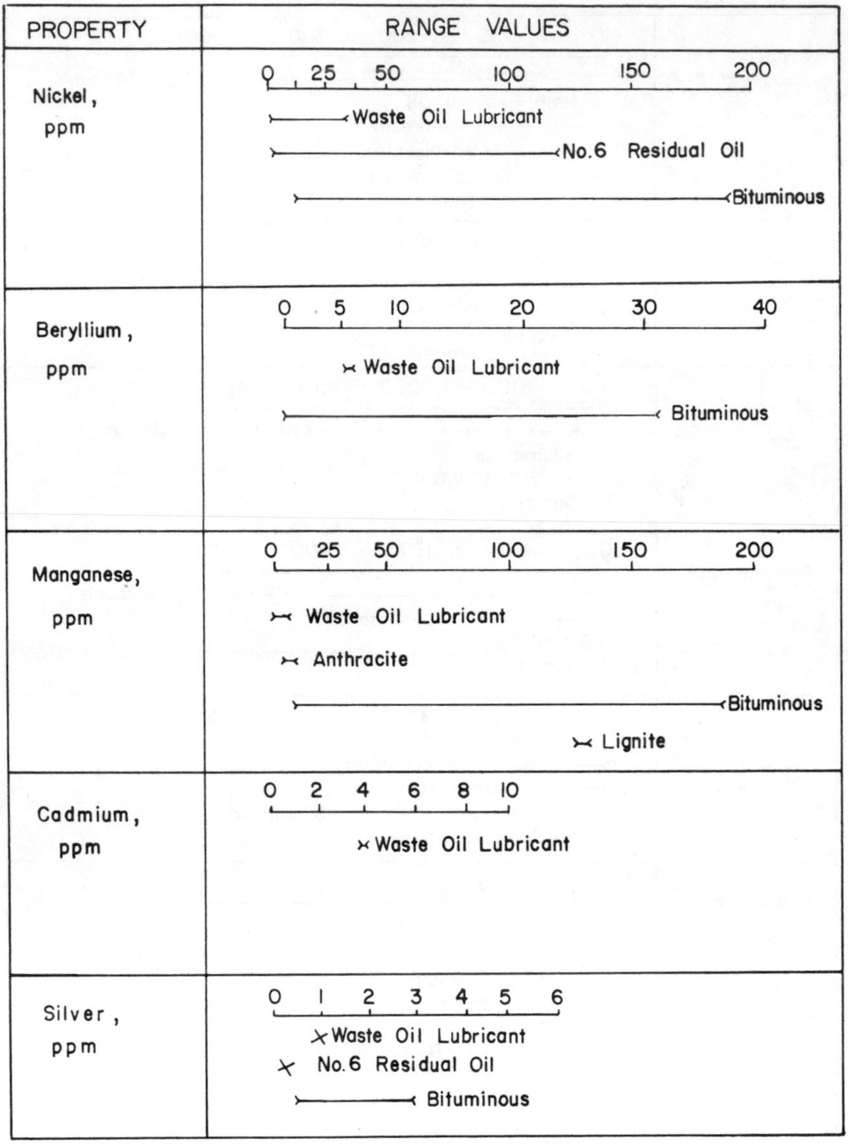

PROPERTY	RANGE VALUES
Nickel, ppm	 0 25 50 100 150 200 ⊢——◁ Waste Oil Lubricant ⊢————————————◁ No.6 Residual Oil ⊢—————————————————◁ Bituminous
Beryllium, ppm	 0 5 10 20 30 40 ⊨⊣ Waste Oil Lubricant ⊢——————————————◁ Bituminous
Manganese, ppm	 0 25 50 100 150 200 ⊢⊣ Waste Oil Lubricant ⊢⊣ Anthracite ⊢——————————————◁ Bituminous ⊢⊣ Lignite
Cadmium, ppm	 0 2 4 6 8 10 ⊢⊣ Waste Oil Lubricant
Silver, ppm	 0 1 2 3 4 5 6 ⊠ Waste Oil Lubricant ⊠ No.6 Residual Oil ⊢———◁ Bituminous

FIG. 47.   (continued)

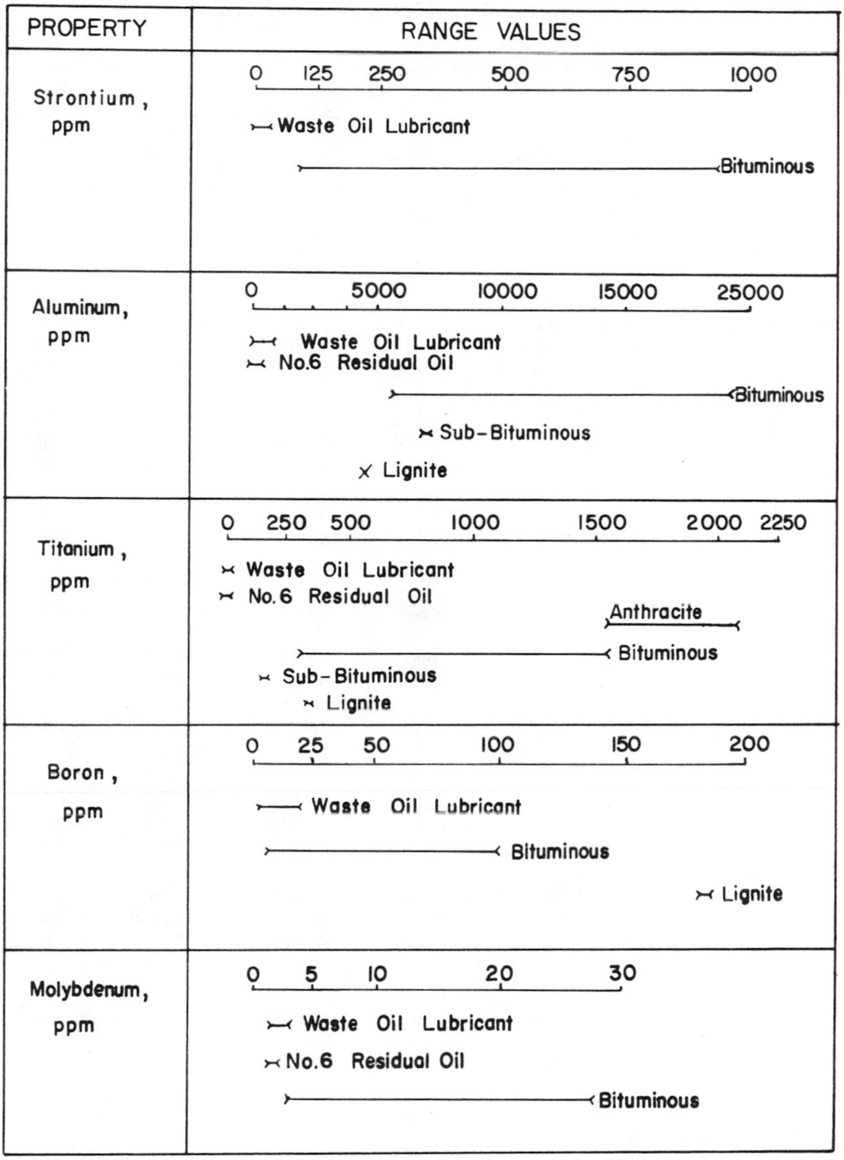

PROPERTY	RANGE VALUES
Strontium, ppm	
Aluminum, ppm	
Titanium, ppm	
Boron, ppm	
Molybdenum, ppm	

FIG. 47.   (continued)

TABLE 59 - SAE VISCOSITY NUMBERS FOR CRANK CASE OILS

SAE Viscosity Number	Viscosity Units	Viscosity Range			
		At 0°F		At 210°F	
		Minimum	Maximum	Minimum	Maximum
5W	Centipoises	NA	1,200	NA	NA
	Centistokes	NA	1,300	NA	NA
	SUS	NA	6,000	NA	NA
10W	Centipoises	1,200	2,400	NA	NA
	Centistokes	1,300	2,600	NA	NA
	SUS	6,000	12,000	NA	NA
20W	Centipoises	2,400	9,600	NA	NA
	Centistokes	2,600	10,500	NA	NA
	SUS	12,000	48,000	NA	NA
20	Centistokes	NA	NA	5.7	9.6
	SUS	NA	NA	45	58
30	Centistokes	NA	NA	9.6	12.9
	SUS	NA	NA	58	70
40	Centistokes	NA	NA	12.9	16.8
	SUS	NA	NA	70	85
50	Centistokes	NA	NA	16.8	22.7
	SUS	NA	NA	85	110
10W30	Centipoises	1,200	2,400	NA	NA
	Centistokes	1,300	2,600	9.6	12.9
	SUS	6,000	12,000	58	70
20W40	Centipoises	2,400	9,600	NA	NA
	Centistokes	2,600	10,500	12.9	16.8
	SUS	12,000	48,000	70	85
20W50	Centipoises	2,400	9,600	NA	NA
	Centistokes	2,600	10,500	16.8	22.7
	SUS	12,000	48,000	85	110

NA   Not Applicable.

[a]Values taken from the 1972 SAE Handbook.

TABLE 60 - SPECIFICATION PERFORMANCE REQUIREMENTS

Test Technique	Test Parameter	Proposed Limit MIL-L-46152[1]	MIL-L-2104C[2]
L-38[3]	Bearing weight loss, maximum	40 mg	50 mg
Sequence IIB	Rust, minimum (10 = clean)	8.9	8.5
	Stuck lifters	None	None
Sequence IIIC	100°F viscosity increase at 40 hr, max	400 percent	NR
	Average engine ratings at 64 hr:		
	Piston varnish, minimum (10 = clean)	9.5	NR
	Oil ring land varnish, minimum (10 = clean)	6.0	NR
	Sludge, minimum (10 = clean)	9.0	NR
	Ring sticking (compression & oil)	None	NR
	Scuffing and wear at 64 hr:		
	Cam or lifter scuffing	None	NR
	Cam plus lifter wear:		
	Average	0.0010 in	NR
	Maximum	0.0020 in	NR
Sequence VC	Total sludge, minimum (10 = clean)	8.5	8.0
	Total varnish, minimum (10 = clean)	8.0	7.0
	Piston skirt varnish, minimum (10 = clean)	8.2	7.5
	Oil ring clogging, maximum	5 percent	5 percent
	Oil screen clogging, maximum	5 percent	5 percent
	Intake tip wear, maximum	RR	RR
	Ring gap increase, maximum	RR	RR
	PCV valve clogging at 18 in Hg, max.	RR	RR
	Ring sticking	None[4]	None[5]
1-II	Top groove filling, maximum	30 percent	NR
	Second groove coverage, maximum	50 percent	NR
	Below second groove	Essentially Clean	NR
1-D	Top groove filling, maximum	NR	75 percent
	Below second groove	NR	Essentially Clean
1-G	Top groove filling, maximum	NR	60 percent
	First land coverage, maximum	NR	50 percent
	Second groove coverage, maximum	NR	30 percent
	Below second groove	NR	Essentially Clean

NR Not required.
RR Rate and report.
1. Includes viscosity grades 10W, 30, 10W30, and 20W40.
2. Includes viscosity grades, 10, 30, 40, and 50.
3. Multigrade oils must stay within their 210°F viscosity grade at the end of 10 hr testing in accordance with FTM 3405 (to be run on stripped drain).
4. Compression and oil.
5. Compression only.

Environmental Impact of Waste Oil

       The detrimental environmental effects of oil pollution are significant and have resulted in local, state, federal and international governmental intervention and the promulgation of many regulations. Sources for the pollutant oil are many including natural seepage, tanker spills, refinery spills, leaks, and the improper disposal of waste oil. Waste oil spills and other improper disposal procedures account for about 15 percent of the total oil entering the waterways of the United States.[64] The impact of waste oil on the environment has not been fully ascertained, but available evidence suggests that it is highly toxic and exhibits many environmentally undesirable traits. The environmental effects of waste oil can be categorized as follows:

Human Health Effects of Waste Oil

       Waste crankcase oil, as a result of additives and products of combustion, contains high concentrations of many of the heavy metals. Along with the 616 million gallons of waste crankcase oil generated annually is an estimated 7 million pounds of lead, 0.5 million pounds of phosphorous, 0.5 million pounds of zinc, and 0.35 million pounds of barium. The high toxicity of the heavy metals (lead, zinc, mercury, etc.) has been well documented for their tendency to accumulate in tissues. Many regulations have been promulgated to reduce human consumption of these heavy metals, and these include the removal of lead from paint, and gasoline, and the elimination of mercury dumping.[65] These metals through waste oil continue to be emitted to the atmosphere and waters and add to the

loading of toxic materials. The lead content of all waste oil generated nationally represents 4 percent of the total lead mined
annually. Waste oil effects on aquatic life directly affect human
life. A recent study presented strong arguments that increased human
consumption of PAH (Polycyclic Aromatic Hydrocarons -- a carcinogenic
substance found in oil) could increase the cancer risk of humans.[66]
A primary source of PAH in human tissues is most likely from the
ingestion of oil-contaminated seafood. There have been indications
that these PAH compounds accumulate and are stored in living tissue.
Waste oil contains many PAH compounds, and can contribute to their
availability to the life cycle.

Effects on Waterfowl and Wildlife

Recent studies[67] regarding the effects of waste oil on
waterfowl and other waterbirds identify waste oil to be as serious
a problem as crude or refined oil. When the bird's outer coating
of feathers becomes covered with oil, its insulating effects are
nullified, allowing the bird to be exposed to the weather elements.
A much more serious problem is the ingestion of waste oil by the
bird trying to clean its feathers. The toxicity of waste oil is high
primarily due to the heavy metal content which adversely affects the
bird's health and egg-laying potential. Ingestion of the waste oil
by the birds also comes through feeding on other waterlife organisms.
Other forms of wildlife are probably less affected by waste oil, but
this is due more to their ability to escape contact with the oil than
by it being less harmful to them. The general impact of waste oil on
each creature will depend on its consumption of oil or oil contaminated

aquatic organisms, its resistance to the oil, and its ability to remove itself from the vicinity of the oil.

Effects on Marine and Freshwater Organisms

There has been limited research conducted on the impact of waste oil on aquatic life, but these do show that heavy concentrations of oil on water surfaces can effect a high fish mortality rate and affect the reproductive cycle. Fish exposed to oil covered water were found to have high concentrations of zinc, lead, and hydrocarbons, which are all constituents of waste oil, in their tissue. As previously noted, these considerations are important in relation to the food chain and ultimate human consumption of these fish and fish products.

Effects of Burning Waste Oil

The combustion of waste oil as an energy source has been a common practice for many years, but recently has been increasing at a rapid pace due to the energy crisis and fuel oil shortages. Nearly one-third of the waste oil used for fuel is not reprocessed or re-refined, and this practice generates potential fire hazards, because waste oil is comprised of additives and combustion by-products. which can be highly flammable, increasing the possibility of explosion and fire. Furthermore, these additives and other sediments in the waste oil can clog oil burner jets and render them inefficient and/or inoperative. The impact on air quality from the combustion of waste oil must also be examined. National air quality standards for lead and other heavy metals have not yet been promulgated and emission standards are vague. However, the U.S. Environmental Protection

Agency has recommended the combustion of waste oil be carried out only when a) highly efficient particulate control systems are used; or b) prior treatment of the waste oil removes the lead content. Studies[68] have been conducted throughout the country and have shown waste oil combustion to effect high lead emissions in submicron particles. Lead oxides in quantities of 400-720 pounds are emitted during the burning of 10,000 gallons of waste oil. This is equivalent to the hourly lead emissions from 18,000 to 32,000 automobiles. Lead availability in submicron particles is a human health hazard since these particles can be inhaled. Physiological dangers detrimental to human health effects result when ambient concentrations of lead particles are about two micrograms per cubic meter and sustained for three or more months. These concentrations of lead compounds have resulted in anemia and miscarriages in humans. The cumulative effects of lead can cause damage to the central nervous system, and recent evidence indicates that automobile exhausts may already be resulting in toxic lead encephalitis (inflammation of the brain) among the urban population.[69] The Federal Energy Office notes that waste oil for fuel oil is poor utilization of resources because of pollutant emissions, and the current needs for lubricating oils and lead.

Effects of Waste Oil on Domestic Sewage Treatment Systems

The performance of a typical sewage treatment system can be affected by oil substances.[70] Waste oil generally enters a sewage treatment plant through a city's storm sewer system in volumes that have still not been established. The oils and resultant oily materials

reduce the effectiveness of the settling agents in a sewage treatment
plant, slowing the removal of solid materials from municipal wastes.
As efficiency of the sewage treatment plant is lowered the oily
materials can pass on through the system and enter waterways.

Effects of Using Waste Oil as a Dust Control Agent

          The final result for waste oil used for ground dust control
is usually water contamination. Some waste oil is washed away by rain
and is conveyed to waterways, and some seeps into groundwaters.
Analyses are generally not available on related impacts. One study
by the Environmental Protection Agency has shown that over a 12 year
period only one percent of the oil placed on a road remained in the
top inch of the road surface.[62] In addition, the composition of the
road surface showed high levels of lead and other toxic materials,
which eventually can join runoff and adversely effect roadside crops,
or at least enter the crops and the human food chain.

Waste Oil as a Fuel

          Waste oil has always been utilized as a fuel. A few years
ago, when fuel shortages were unheard of it served as a cheap
supplement to Number 4 Fuel Oil and cost around 4-6 cents per gallon.[56]
As shortages in fuel surfaced, it increased and reached 20 cents per
gallon in December of 1973.[61] At these prices the processing of
waste oil was uneconomical and in fact, waste oil, as a fuel, received
little or no processing.[71] Generally, the oil was just processed to
remove water, and suspended solids and any solubilized materials were
burned along with the oil in the burner.

The burning of waste oil has little or no effect on the
burner and heat transfer surfaces of the furnace as long as there are
negligible amounts of dissolved metals, which is usually the case for
industrial lubricants. Most lubricating oils are desirable fuels
because they have heating values similar to virgin oils or 120,000
million BTU per gallon and have low sulfur contents, making them
attractive for their low sulfur dioxide emissions.

Waste oil from automotive and similar processes contains
high percentages of solubilized metals. Upon combustion of the fuel
these metals tend to clog the burner nozzles, and therefore, the
waste oil must be diluted. Typically, waste oil is burned along with
conventional fuels. in quantities of 5 percent or less. As a supplementary fuel, waste oil does not present a clogging problem, but
corrosion and fouling of boiler tubes and undesirable stack emissions
become factors to be considered. The American Petroleum Institute
conducted studies at major refineries and utilities and found stack
emissions to contain essentially all the lead which was fed into the
furnace.[67] Stack emissions of lead were as low as 3 percent but more
typically were 30 percent of the lead content of the feed with the
remainder fouling tubes.[72] Operating conditions including soot blowing
techniques determined the fraction that remained on the tubes and the
fraction emitted.

As noted, the recent rise in price has made the use of
waste oils as a fuel economically attractive. The value of these
oils to a utility in terms of heating value was about 13¢ per gallon,

based on fuel valued at $1.00 per million BTU. This compared to a

"crude" waste oil cost for re-refiners of between 2 and 4 cents per

gallon during the summer of 1973. While they have become economically

attractive, the use of waste oils as fuel creates environmental problems

that must be satisfied. Their use necessitates the utilization of air

pollution control equipment capable of removing all lead and other

metals from the flue gas prior to atmospheric emission.

        To reduce the environmental impact of the combustion of

waste oil the processes available include: (1)   fuel blending,

(2)   waste oil pretreatment, and/or   (3)   use of more efficient

emission control devices. The latter process does not reduce the

operating problems associated with burning raw waste oil, such as

jet clogging, corrosion, water content, etc., and should probably

only be used as a supplement to fuel blending or waste oil pre-

treatment.

        Several other waste oil burning tests have been conducted

to evaluate the feasibility of utilizing waste oil to raise the BTU

content of incinerators destined for municipal solid wastes.[73]. These

studies indicated that the fuel rate required to fire the wastes could

be lowered if the solid wastes were presoaked with waste oil. How-

ever, the combustion of the oil-soaked product effected high parti-

culate emissions and a black opaque plume. Similarly, coal was pro-

cessed with waste oil prior to combustion, and because coal burning

boilers already require extensive air emission control devices new

pollution abatement costs were not necessary.[74] Another study demon-

strated the use of lead concentrated waste oil as fuel for a secondary

lead reverberatory furnace. These furnaces normally have extensive
air pollution control devices eliminating the need for additional
control equipment. The economical advantages of this process are
twofold: (1) an inexpensive fuel source, and (2) the recovery of
lead.[75]

The technical and economical feasibility of utilizing waste
oil as a supplement to conventional fuels is dependent on many parameters including: (1) additional storage and handling facilities may
become necessary, (2) the blending compatability of the waste oil
with other fuels, (3) any combustion impacts created by such a
utilization, and (4) other impacts resulting from the introduction
of foreign materials into a system via the waste oil. Tables 61,62,63
characterize waste oil/distillate oil, waste oil/residual oil,
and waste oil/bituminous coal blends, respectively.

Several factors affect fuel blending possibilities.
While the specific gravity of the fuel is lowered, the water,
solids, and ash forming components are only diluted. The heavy
metals are still emitted to the environment and tend to become
concentrated within process equipment to promote clogging, corrosion, etc. Pretreatment to remove the detrimental components
before combustion seems to be the obvious solution and Table 64
shows the potential impact reduction alternatives for utilizing
pretreated waste oil as opposed to untreated waste oil as a fuel.

The function of a primary pretreatment process is to
remove volatiles and coarse solids in order to minimize the abrasive
wear of nozzles and valves as well as to produce a fuel with a

TABLE 61 - CHARACTERIZATION OF WASTE OIL/DISTILLATE BLENDS[a][75]

Property	Value					
	0% Waste Oil	1% Waste Oil	5% Waste Oil	10% Waste Oil	25% Waste Oil	100% Waste Oil
Gravity, °API @ 60°F	37.8	37.7	37.1	36.4	34.4	24.0
Viscosity, Centistokes	3.0	3.0	4.0	5.0	11.0	99.0
Pour Point, °F	-12.5	-12.7	-13.6	-14.8	-18.1	-35
Flash Point, °F	165	166.3	171.5	178.0	197.5	295
Heating Value, BTU/lb	19,020	18,994	18,891	18,762	18,374	16,436
BS&W, Vol. %	0.05	0.16	0.60	1.15	2.79	11.0
Sulfur, Wt. %	0.310	0.311	0.316	0.322	0.340	0.430
Ash, Wt. %	0.0025	0.022	0.098	0.193	0.479	1.91

[a]The properties of these blends are assumed to be linearly related to their constituents' properties except for viscosity which was calculated using the Kendall-Monroe equation.

TABLE 62 - CHARACTERIZATION OF WASTE OIL/RESIDUAL OIL BLENDS[a] [75]

Property	Value					
	0% Waste Oil	1% Waste Oil	5% Waste Oil	10% Waste Oil	25% Waste Oil	100% Waste Oil
Gravity, °API @ 60°F	13.2	13.3	13.7	14.3	15.9	24.0
Viscosity, Centistokes	379	365	345	338	288	99.0
Pour Point, °F	52.5	51.6	48.1	43.8	30.6	-35
Flash Point, °F	210	211	214	219	231	295
Heating Values, BTU/lb	18,945	18,920	18,820	18,694	18,318	16,436
BS&W, Vol. %	1.00	1.10	1.50	2.00	3.50	11.0
Sulfur, Wt. %	2.15	2.13	2.06	1.98	1.72	0.43
Ash, Wt. %	0.25	0.27	0.33	0.42	0.67	1.91
Silicon, ppm	86.1	89.7	104	122	175	443
Calcium, ppm	47.9	65.9	138.0	228	498	1,850
Sodium, ppm	240.5	239.7	236	232	220	158.0
Iron, ppm	120.3	129	166	211	346	1,025
Magnesium, ppm	14.2	19.6	41.4	68.7	150.4	559
Lead, ppm	2.9	63	303	603	1,502	6,000
Vanadium, ppm	190.5	188.8	182	171.5	148.1	21
Copper, ppm	0.5	2.3	9.3	18.2	44.6	177
Chromium, ppm	13.7	13.9	14.5	15.2	17.5	29
Nickel, ppm	60.5	60.1	58.3	56.1	49.5	16.5
Silver, ppm	0.3	.31	.34	.37	.48	1
Aluminum, ppm	109.8	113	125	139	184	405
Titanium	5.5	5.6	6.1	6.7	8.5	17.5
Molybdenum, ppm	2.3	2.30	2.31	2.32	2.35	2.5

[a]The properties of these blends are assumed to be linearly related to their constituents' properties except for viscosity which was calculated using the Kendall-Monroe equation.

227

TABLE 63 – CHARACTERIZATION OF WASTE OIL/BITUMINOUS COAL BLENDS[a][75]

Property	Value					
	0% Waste Oil	1% Waste Oil	5% Waste Oil	10% Waste Oil	25% Waste Oil	100% Waste Oil
Heating Value, BTU/lb.	12,486	12,526	12,684	12,881	13,474	16,436
Sulfur, Wt. %	2.75	2.73	2.63	2.52	2.17	0.43
Ash, Wt. %	10.5	10.4	10.1	9.64	8.35	1.91
Silicon, ppm	26,650	26,388	25,340	24,029	20,098	443
Calcium, ppm	7,768	7,709	7,472	7,176	6,289	1,850
Sodium, ppm	469	466	453	438	391	158.0
Iron, ppm	14,467	14,333	13,795	13,123	11,106	1,025
Magnesium, ppm	1,362	1,354	1,322	1,282	1,161	559
Lead, ppm	78.5	138	375	671	1,559	6,000
Vanadium, -ppm	30	29.9	29.6	29.1	27.8	21
Copper, ppm	64	65.1	69.7	75.3	92.3	177
Barium, ppm·	258	265	295	333	445	1,005
Zinc, ppm	123	138	199	276	505	1,650
Phosphorus, ppm	30	42	91	152	335	1,250
Tin, ppm	225.2	223.5	216.9	208.5	183.5	58.5
Chromium, ppm	24	24.1	24.3	24.5	25.3	29
Nickel, ppm	101	100	97	93	80	16.5
Beryllium, ppm	15.6	15.5	15.1	14.6	13.2	6
Manganese, ppm	101	100	96	92	78	7.5
Silver, ppm	1.7	1.69	1.67	1.63	1.53	1
Strontium, ppm	515	510	490	466	391	20
Aluminum, ppm	12,503	12,382	11,898	11,293	9,479	405
Titanium, ppm	945	936	899	852	713	17.5
Boron, ppm	54.7	54.3	52.5	50.4	43.9	11.5
Molybdenum, ppm	15.6	15.5	14.9	14.3	12.3	2.5

[a]The properties of these blends are assumed to be linearly related to their constituents' properties.

TABLE 64 - POTENTIAL IMPACTS AND IMPACT REDUCTION ALTERNATIVES OF UNTREATED WASTE OIL UTILIZATION AS A FUEL[75]

Property	Potential Impacts	Impact Reduction Alternatives
Specific Gravity	Formation of concentration gradients in combined storage tanks with distillate oils	• Storage in tanks that accomplish mixing via convectional heating coils. • Separate storage with blending just prior to combustion.
Water	Fuel line freezing	• Use with heated fuel lines. • Removal of water prior to use (low-level pretreatment).
	Burner flameout	• Use with auxiliary torch to sustain burner flame.
	Inconsistent heating value	• Use for temperature insensitive application. • Removal of water prior to use (low-level pretreatment).
Coarse Solids	Sludge buildup in storage tank to point of drawoff	• Storage in tanks with bottom sludge removal drains. • Use with dispersant emulsifiers to keep sludge in suspension. • Removal of sludge prior to use (low-level pretreatment).
	Line strainer fouling	• Removal of sludge prior to use (low-level pretreatment).
	Abrasion of positive displacement pump seals	• Separate waste oil storage plus transport prior to blending with hardened impeller centrifugal pumps. • Removal of sludge prior to use (low-level pretreatment).
	Abrasion of burner nozzles	• Use with wide orifice hardened nozzles. • Removal of sludge prior to use (low-level pretreatment).
Ash Forming Materials	Health hazard to boiler cleaning crew	• Use of respirators during cleaning. • Removal of ash forming materials prior to use (high-level pretreatment).
	Scaling and corrosion of heat transfer surfaces	• Use in direct-fired furnaces. • Removal of ash forming materials prior to use (high-level pretreatment).
	Hazardous emissions	• Use with efficient particulate emission control equipment. • Removal of ash forming materials prior to use (high-level pretreatment).
	Ash disposal problems	• Removal of ash forming materials prior to use (high-level pretreatment).

consistent heating valve. Primary pretreatment will keep operating
and maintenance costs of feedstock and burner systems of a waste oil
facility comparable to those incurred at a conventional fuel facility.
However, primary pretreatment has little affect on the metallic constituents of waste oil and subsequently does not significantly reduce
the fouling and corrosion of boiler heat exchange surfaces or the
emission of metallic contaminants that would result from waste oil
combustion. In order to obtain significant metallic contaminant
removal, secondary pretreatment techniques must be utilized.

Filtration is the most common process for effecting liquid-
solid separations. All petroleum fuel oil handling facilities utilize
this unit operation, some in the form of simple strainers. Self-
cleaning edge type filters are often used but are only effective
for separating particles greater than 40 microns. They remove
abrasive particles but have no great affect on the metal content
of the waste oil. The effective separation of micron-sized particulates from the waste oil can only be achieved by fine mesh filters
which have high capital and operating costs. Like most other processes, the cost of filtering waste oil varies inversely with the
size of the particles. Filtration can be effective after chemical
treatment causes flocculation of the particles. Therefore, filtration is frequently used after acid/clay treatments effecting high
metal content sludges which must then be treated.

Settling is the simplest process for removing the portion
of the bulk solids and water (BS&W) not held in suspension by the

oil and its dispersants. Separation is effected by gravity. The
rate of settling can be increased by heating the oil and lowering
its viscosity. However, since most of the carbon, metallic and
other particles that exist in waste oil are less than 1 micron in
size this procedure is not very effective at lowering the contaminant
levels. Table 65 shows the effectiveness of particle settling with
respect to size. It indicates the calculated settling rates of
particles (3.0 g/cc, Sp. Gr.) in a typical waste oil at 100°F.

At 200°F the settling rates of the particles would be 5 to
10 times greater than those shown in Table 65 but still be too low to
be effective for all but the largest particles. Although settling
is largely ineffective in separating fine contaminants, it does
remove coarse grit and free water, and is normally the first step
of most treatment processes. The transfer of waste oil from storage
tanks always results in withdrawal of the oil from a level above the

TABLE 65 - SETTLING OF PARTICLES IN 100°F WASTE OIL[53]

Particle Diameter (μm)	Time to Settle Through 1 Ft.
0.1	40 years
1.0	160 days
5.0	6.5 days
10	1.6 days
100	25 minutes

BS&W sediments level in the tank. The BS&W that accumulates is periodically removed from the bottom of the tank. This procedure is the simplest form of contaminant removal by settling. More rigorous treatments would involve controlled settling at high temperatures or with additives to reduce oil viscosity. Figure 48 is the process flow plan of a typical settling pretreatment system.

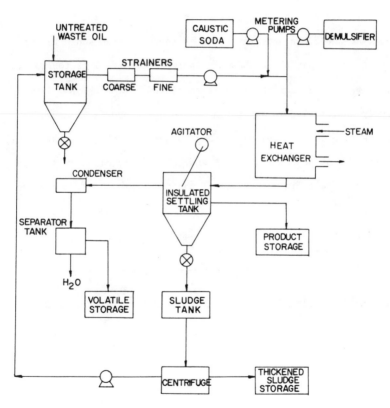

FIG. 48.   Settling pre-treatment.

Centrifuging can greatly increase settling rates. Commercial units generate forces many thousands of times that of gravity, proportionately reducing settling times. For a separating force of 10,000 times that of gravity the 1-micron particle shown in Table 65 will settle 1 foot in about 25 minutes. Commercial units are generally ineffective for 1 micron and smaller particles, but can remove, with high efficiencies, particles and water droplets 3 to 5 microns in diameter. The waste oil is normally heated prior to centrifugation to improve fluid flow and particle movement. Another technique used to lower the viscosity of waste oil is to add to the oil an oil soluble solvent such as naptha or toluene. Demulsifiers are sometimes used in conjunction with centrifuging to effect coagulation of the suspended particles and water droplets and increase separation. Generally, centrifuging is a technically feasible method for processing large volumes of waste oil, but for limited separation. Ash and metallic contaminant levels are not appreciably lowered because of their fine particle size. Figure 49 is the process flow diagram of a typical centrifugation pretreatment system.

Demulsification, followed by a solid-liquid separation method, is a technique used by some waste oil processors. One waste oil reprocessor has succeeded in lowering the BS&W content of the waste oil to a maximum of 1.5 percent through addition of a demulsifier followed by centrifuging.[76] Demulsifiers are added to the waste oil to counteract the effects of still active emulsifiers

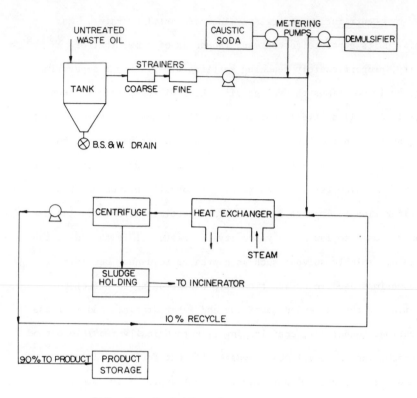

FIG. 49.   Centrifugation pre-treatment.

which keep water and solids in suspension. Demulsification is highly
efficient in affecting the removal of water. Solids are separated with
more difficulty. Surfactants have been used to wet the surfaces of
these solids under elevated temperatures (200°F). This treatment
only effects the removal of relatively large particles.

        The removal of the volatile components of waste oil is
effectively accomplished by thermal processes which vaporize the
low boiling contaminants. Flash distillation at atmospheric or
reduced pressures is a common process for the removal of volatiles.

Some re-refiners utilize the thermal inputs of settling or centrifugation operations to drive off most of the water and fuels with
low boiling points. Volatiles are condensed and water is separated
from immiscible organic liquids by decantation processes.

Secondary pretreatment techniques such as ultrafiltration
and membrane dialysis are theoretically capable of obtaining fine
particle separations but are generally impractical because of the
high solids content of waste oil. Other treatment methods or combination of methods such as demulsification, electrolytic deposition,
flocculation, and coagulation can also separate fine particles. Yet
none appear presently to be commercially available to separate the
many metallic constituents from a product as variable as waste crankcase oil. Metals' separation by such treatment methods or combination
of methods would probably be possible only after extensive laboratory
and pilot plant development to establish feasibility. However,
several treatments can be used to effect the removal of metallic
contaminants. The most common of these are (1) acid/clay treatment,
(2) solvent extraction and, (3) vacuum distillation. These methods
are the major re-refining processes and are normally utilized to
produce lube oil stock rather than fuel oil because greater profits
are realized from lube oil. They will be discussed in the following
section on re-refining.

Waste Oil Re-refining

Presently there are but a few operating re-refiners. As a
result of economic and technical pressures the re-refining industry
has been reduced over the last few years from over 150 to less than

40 companies.[77]   The major problems encountered by the industry are
environmental (waste disposal), operational (obtaining sufficient
waste oil feedstock), and political (price incentives given to virgin
oil refiners and the recycled oil labeling laws).

Presently, most re-refiners in the United States use the
acid/clay process. The process flow diagram for the typical acid/clay
re-refining of waste automotive oils is shown in Figure 50. The process to re-refine diesel engine waste oils differs only in process
operating conditions.

The collection trucks unload their waste contents into a
partially submerged tank. This receiving tank is fitted with grids
and screens to remove the larger contaminants normally found in the
waste materials. The receiving tank is large enough to accept the
entire truck load and permit any free water to settle. The oil is
then decanted and transferred to feed storage tanks. The water layer
passes to a skimmer and then to a wastewater treatment facility.
Handling of the raw material is extremely important for continuous
operation of the process.

A typical waste oil analysis in the storage tanks would be:

Water	3.5% by volume
Naptha	7.0% by volume
Oil, etc.	89.5% by volume

The waste oil passes through a steam heat exchanger to the
flash dehydrator which operates at 300°F and atmospheric pressure.
The steam/oil overhead is condensed and separated. The oil is trans-

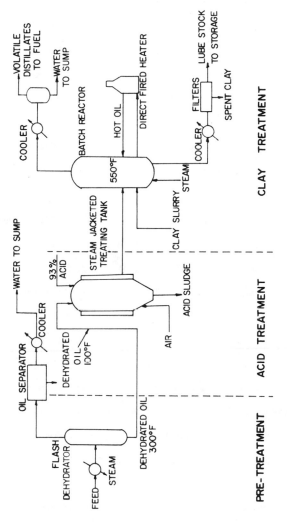

FIG. 50. Re-refining by an acid clay process.

ferred to light end storage to be used as fuel, and the water to the
wastewater treatment facility.   The dehydrated oil is sometimes
pretreated before acid processing, but usually it is just stored and
cooled in dry oil tanks after the heat exchanger.   It can stay in these
storage tanks for 2 to 4 days before appreciable amounts of moisture
accumulate which tend to increase the requirements of acid in the next
process.   After about 48 hours in storage the oil has cooled to approximately 100°F.

This waste oil is then pumped to one of several acid treating
units.   These units are steam jacketed and aerated for mixing.   The
temperature of the reactor is maintained at 100°F and 93 percent
sulfuric acid solution is added in quantities of four to six percent
by volume.   The oxidized constituents of the oil are usually coagulated
within 24 hours, but depending on the waste oil longer residence times
may be required.   The acid sludge, containing oil contaminants and
ash, separates from the oil and is withdrawn from the reactor bottom.
The disposal of the acid sludge disposal which is usually landfilled
is one major problem inherent in this process.

The acid treated dehydrated oil is then passed on to the
steam stripping-clay treatment process.   The clay treater is equipped
with an overhead condensing unit and a direct fired heater through
which the oil is circulated.   A sparger is used for the direct introduction of steam.

The temperature of the batch is raised to 550-600°F by circulating the acid treated waste oil through the clay treatment heater
and the introduction of live steam into the batch.   This stripping

operation is to remove the remaining light fuel fractions and.any
mercaptans which may be present and normally continues for 12-15 hours.
The steam-stripped materials are condensed and passed through a separation process. The water fraction is treated in the wastewater
treatment system and the oil fraction used as plant fuel. The remaining waste oil is treated as follows: The heat to the clay
treatment heater is discontinued and a fraction of the fuel oil is
diverted to the clay slurry tank where the oil temperature is lowered
to approximately 400°F. The clay which is normally 50 percent activated
clay and 50 percent diatomaceous earth (200-250 mesh) is mixed into
the circulating oil. Clay is added at a rate of approximately 0.4
pound per gallon of oil. The clay adsorbs color bodies as well
as colloidal carbon.

The clay mixture passes through a plate and frame filter
press and sometimes a second filter in series. The clarified oil
is then stored, and the necessary additives are blended into the
stock.

The residue or filter cake is a mixture of clay, impurities, and oil, and is uneconomical to separate and recover in
small plants. Therefore, it is usually discarded at landfills.
Alternate disposal methods are being sought because regulations are
being promulgated to prohibit these procedures.   The EPA is
currently investigating treatment methods of acid sludge.

Acid/clay re-refining can generate many odors. They may
be generated by storage tanks, processing vessels, wastewater treatment facilities, acid sludge, and oil spills. In some re-refining

operations, odors can be controlled adequately by sealing open vessels and tanks, general maintenance, and through burning the vapors given off by process vessels with the normal fuel. Caustic scrubbers have been used for odor control in some plants.

The wastewater facilities vary from plant to plant. They are dependent on cooling water and vacuum equipment, water runoff problems, land availability, water contamination of feed stocks, governmental regulations, and availability of local treatment facilities. A typical installation includes an API separator with oil skimming, a pH control device, some water recycle, and discharge to a sewage plant. Sewage treatment plants will normally accept water with oil contents up to about 100 ppm, a quality level relatively easy to meet. Little data is available on the other waste oil process wastewater characteristics.

The final oil product yields for acid/clay treatment have been reported to be from 45 to 75 percent of initial throughput. These are dependent upon operating conditions of the plant and waste oil composition, such as water, sludge, ash, and gasoline contents. For many of the common waste oils (i.e. 3.5% H_2O , 7.0% naptha), a greater than 70 percent output is possible with efficient operation.

The oil produced by the acid/clay process can be considered a solvent of neutral blending stock, with properties approximately equal to SAE 20 lubricating oil. The SUS (Saybolt Universal Second) viscosity is generally between 55 and 58 at 210°F. This oil can be processed to a finished lube by the re-refiner or sold directly to

blending operators. The viscosity of the derived oil can be increased
by the addition of virgin oils, or by the addition of polyisobutylene.
Conventional additive packages are used when high performance
specifications are to be met.

The Vacuum Distillation Process, as shown in Figure 51, overcomes
the serious acid sludge waste disposal problem associated with acid/
clay treatment. Some re-refiners are now switching to such a process.
However, when the process utilizes clay contacting as tertiary treatment these disposal problem advantages are negated. Before vacuum
distillation, the waste oil is usually pretreated. The collected
"crude" waste oils are received as before and passed through screens
to ensure that no extraneous sediments enter the process stream.
The waste oil is heated to 300°F in a direct heater which uses as
fuel the light hydrocarbons that evolve during processing. The
waste oil passes in the flash tower which operates at atmospheric
pressure and 300°F. The oil/water vapor mixture on top is condensed
and withdrawn to an oil decanter. The water fraction is separated
and removed to the wastewater disposal system. The oil fraction is
used as fuel in the plant.

The bottoms from the flash tower are cooled in a heat
exchanger to a temperature of approximately 100°F. Light oil, having
a boiling range of 150-250°F, is added into the dehydrated oil stream
in quantities of approximately 20 percent by oil volume. A small
amount of caustic 0.2-2.0 percent is also added depending on feedstock
characteristics. The light oil and caustic are added to break the

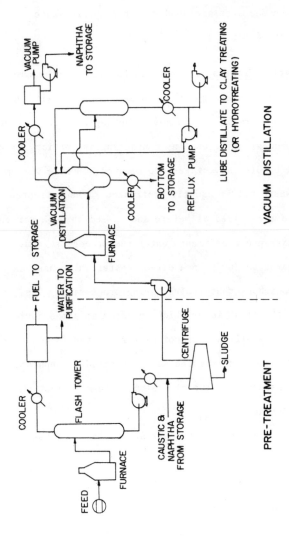

FIG. 51. Vacuum distillation of crankcase waste oil.

oil-water emulsion and precipitate solids. These materials are separated by centrifugation. The sludge from the centrifuge can be disposed of by itself at a landfill, or it can be mixed with the distillate bottoms as described below. The naptha/caustic/centrifuge pretreatment process eliminates some of the materials which can cause fouling and corrosion in the vacuum distillation process equipment.

In the vacuum distillation tower the centrifuged oil is pumped through a direct fired heater. This furnace heats the oil to about 700°F. The vacuum columns distill the oil in a vacuum of 27 inches of mercury. The overhead naptha is condensed and used as fuel in the plant. The bottoms of the distillate column, which contain nearly the entire ash content of the feed, are cooled and used as fuel, or blended into asphalt, or stored in a lagoon. The remaining oil can be sent on to clay process for tertiary treatment as a lube blending stock.

This clay treatment is optional and is similar to the previously discussed acid/clay process except that steam stripping is unnecessary and clay requirements reduced to about 0.125 lbs. per gallon of oil. The filter cake clay residue is disposed of in landfills.

The final oil ouput of this process is about 70 percent of the input waste oil, comparable to the most efficient acid/clay treatment operation.

The oil quality produced by the distillation process is comparable to that produced by acid/clay treating. However, it is generally believed that continued vacuum distillation column process-

ing instead of clay treatment will effect oils with high viscosities.
Clay treatment only improves color and the pH, and reduces the oxygen
and nitrogen contents.

Odors and wastewater don't present any more difficulties
with this process than with acid/clay treatment.[53]

The Solvent Extraction/Acid/Clay Process

The solvent extraction/acid/clay process is a relatively
new development in the re-refining of waste lubricating oil.[78,79]
There have been pilot plants in the United States studying this
process but no plants are presently in operation. A 9 million gallon
per year plant is in operation in Italy based on a process developed
by Institut Francais du Petrole (IFP).[80] Similar processes have been
developed and patented but have not been commercialized in the United
States.[81]

The process is based on the use of propane which selectively
extracts the base oil from the additives and impurities. Figure 52
is a process flow diagram of the propane extraction process. The
propane and dissolved oil mixture is removed from the top of the
extractor. The high boiling, dark colored asphaltic and oxidized
hydrocarbons and suspended solids residue is withdrawn from the
bottom of the unit. The bottoms are then utilized along with conventional fuel oil for plant fuel. Where this is impossible landfill
techniques are used. Propane is flashed from the oil and recycled.
The flow process is as follows:

. Thermal dehydration

. Precipitation and solvent extraction

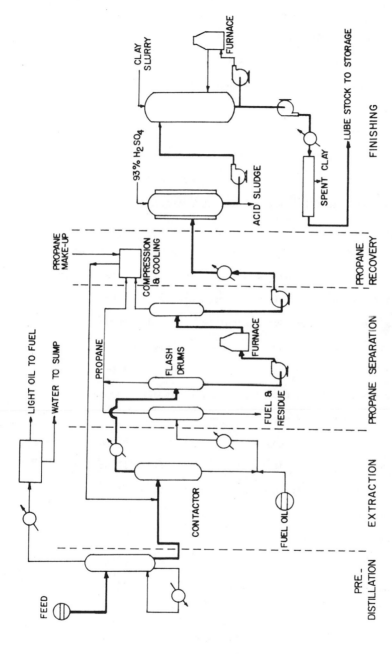

FIG. 52.   Re-refining by a propane extraction process.

. Vacuum distillation

. Acid treatment

. Clay treatment and filtration

The incoming waste oil is unloaded into a receiving tank from
collection trucks. Initially the waste oil is pumped through the
steam heat exchanger to the flash dehydrator operating at about 300°F
and atmospheric pressure. The overhead is condensed and drained into
the oil separator. The water fraction passes to the wastewater disposal system and the oil fraction is either stored or processed
immediately in the solvent extractor.

The remaining oil is pumped to the precipitation tower
(solvent extractor) via a heat exchanger. Heated propane is introduced into the tower at a height of about one-third up from the
bottom and the oil is introduced about one-third down from the top.
The solvent extractor operates at about 500 psig and elevated
temperatures. The propane-oil solution (oil dissolved in the
propane) rises and precipitate flows to the extractor bottom, due
to specific gravity differences.

To obtain very high quality lube oil, the propane-to-waste
oil ratio would be approximately 20:1. This ratio will be dependent
on each waste oil composition. The lowest feasible propane-to-waste
oil ratio would be 1:1, yielding poor quality oil. Present processes
operate with ratios of about 15:1.

The precipitate residue then passes from the extractor. A
small amount of fuel oil is added to the residue pipeline to assist

flow from the unit. Removal of residue from the extractor is controlled by liquid level switches. The fuel oil-residue mixture is
stored for use as fuel for the direct fired heaters (furnace) or for
future disposal. The propane-oil solution is initially flashed
through a pressure reducing valve and then through a solvent flash
drum. It is usual to use a two-stage flash to separate the propane
and oil. The first stage operates at 250°F and 250 psig. The
propane gas is then liquified and recycled.

The separated lube oil passes on to acid/clay treatment as
previously described. The process differs only in the amount of
acid and clay used. This process requires only about 2 percent of
93 percent sulfuric acid by oil volume, compared with the 4-6 percent
required for the acid/clay process. Clay treatment requirements are
only about 0.15 lbs. per gallon at 300°F and then filtration. The
lube oil quality is in many respects superior to the acid/clay
product in terms of color and color stability, and viscosity.

Although the requirements of acid and clay in the IFP
process are greatly reduced, a disposal problem still exists. Acid
sludge analyses are not available, but the metals content including
lead is lower than the product of the acid/clay process. Most of the
metals and other contaminants remain in the fuel oil-residue mixture.
Therefore, the use of this mixture for fuel can result in many detrimental environmental effects, unless large capital investments in air
pollution control equipment are made and large maintenance costs for
furnace corrosion and boiler tube fouling are considered.

The Distillation/Hydrogen Treating Process

The distillation/hydrogen treating process is similar to distillation/clay treating except for the final process. Figure 53 is a process flow diagram. This treatment process is widely used in the petroleum industry, but has made no inroads into the waste oil industry. However, two European installations currently being constructed have plans to combine hydrogen treating with previously discussed IFP propane extraction process.

A pretreatment process usually precedes vacuum distillation to reduce fouling and corrosion problems. The distillate (sidestream) from the vacuum distillation column is heated in a furnace before being mixed with a standard commercial hydrotreating catalyst in a fixed bed. The hydrogen reacts with the oxygen and nitrogen containing impurities and becomes unsaturated.

The pressure is reduced in two flash drums in series to effect hydrogen gas recovery which is then recycled. The purified oil is used to preheat the incoming feed and then injected into a stripping column to remove the small amount of volatiles which may have formed. The purified product leaving the stripper can be used to preheat vacuum distillation feed before final cooling and storage.

Recent work has shown that the hydrogen treated distillate can match typical properties of certain neutral lube blending stock.[82] Hydrotreatment operating conditions used in this work were 650 psig at 650°F with 800 standard cubic feet of hydrogen recycled per barrel of waste oil, and a space velocity of 1.0 v/v/hr.

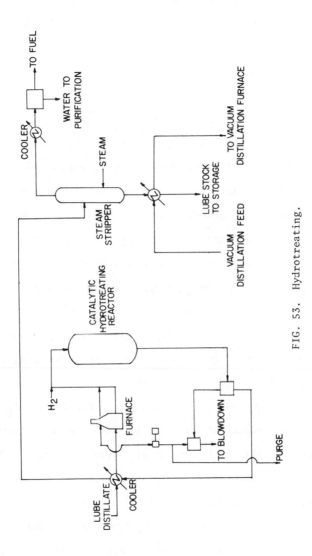

FIG. 53. Hydrotreating.

The distillation bottoms which contain nearly all of the contaminants are deposited in landfills. However, in conjunction with current distillation/hydrogen treating process developement, plans are being made to treat these high lead content bottoms into a secondary lead smelting operation. Should this prove successful, distillation/hydrogen treatment would be the first re-refining process available without a solid waste disposal problem. The wastewater generated is similar to other re-refining processes and can be overcome by conventional design. A scrubber may be required to remove impurities from the hydrogen purge emissions and other minor gaseous discharges. Additional work to improve and prolong catalyst life and reduce hydrogen consumption will be necessary before the commercialization of this process.

The function and product of each of the waste oil re-refining processes are shown in Table 66. Treatment trains which incorporate several of these processes will be able to remove all of the impurities and produce a high quality lube stock.

The economics of waste crankcase oil reprocessing have been studied and detailed for a typical 5 million gallon per year plant. The economics of such a plant are summarized in Table 67. For reprocessing plants of this size to produce a lube blending stock, the distillation/clay treatment is probably the lowest cost process. The economics are such that it is often difficult to differentiate costs between straight distillation with clay treatment, distillation with hydrotreating, and extraction followed by acid/clay treatment processes. However, the acid/clay process, in large plants, appears to require less capital and maintenance investments. In

TABLE 66 - POTENTIAL WASTE OIL RE-REFINING UNIT PROCESSES[53]

Unit Process	Function	Product
A. Settling	Removal of gross water, coarse solid solids and other materials heavier than the oil	"Clarified" oil for further processing
B. Centrifugation	Removal of water, solids and other materials heavier than the oil	"Clarified" oil for further processing
C. Flash Distillation	Removal of light ends, naphtha and water	Naphthas and water-free oil product or oil for further processing
D. Chemical Treatment (Caustic Wash, Demulsification, Oxidation and Agglomeration)	Removal of acidic compounds, additives and contaminants stabilized in solution and suspension	Purified oil for further processing
E. Sulfuric Acid Treatment	Removal of additives and contaminants stabilized in solution and suspension	Purified oil for further processing
F. Fractionation	Separation of heavy contaminants plus splitting of oil into various fractions	Various oil cuts for further processing including stabilization
G. Hydrofining	Reduction of nitrogen, sulfur and oxygen contaminants with hydrogen	Purified base lube oil stock
H. Solvent Extraction	Selective extraction of oil and contaminants	Fuel oil stock or refined base lube stock
I. Adsorbent Treatment	Adsorption of metallic sulfuretted, chlorinated and other additives plus odor and color bodies	Purified oil slurried with contaminated adsorbent
J. Filtration	Removes suspended and settleable solids from oil	Oil or base oil stock

TABLE 67 - WASTE LUBE OIL RE-REFINING PROCESS COMPARISON[53]

(5 Million Gal/Yr WASTE OIL)

Process	Primary Product	Yield, % To Lube Stock	Investment 1000S (1)(2)	Operating Cost ¢/Gal (1)(3)	Primary Wastes
Acid/Clay	Lube Blending Stock	72	1,200	18.2	Acid sludge Spent Clay
Solvent Extraction/ Acid/Clay	Lube Blending Stock	84	1,400	15.5	High ash rafinate Acid sludge, Acid/clay Spent sludge, Acid/clay
Distillation/Clay	Lube Blending Stock	76	1,200	13.8	Spent clay High ash bottoms
Distillation/Hydrotreat	Lube Blending Stock	76	1,400	15.4	High ash bottoms

(1) December 1973

(2) 250 day operation

(3) Exclusion of feed, blending, packaging and additives

smaller facilities, where the disposal of the acid and clay is
feasible the acid/clay treatment would probably still be economically
advantageous because of the smaller capital investment costs.
Specifications for re-refined lubricating oil produced by the acid/
clay process are shown in Table 68.[77] Properties of re-refined
waste oil from various other processes are shown in Table 69.[78]

TABLE 68 - SPECIFICATIONS FOR RE-REFINED OIL[77]

	Viscosity ssu				Pour	%
	100°F	210°F	Flash	V.I.	Point	Ash
Motor Oils						
SAE20-5A Grade	320	54.4	420	104	-20°F	0.0
SAE20W-5C Grade	320	55.0	420	104	-25	1.3
SAE30-SF Grade	540	65.5	450	96	-10	1.3
SAE10W-40-SE Grade	322	74.0	395	153	-45	1.3
SAE30-Series 3	540	65.5	450	96	-10	1.8
SAE10-SD Grade	185	46.5	385	110	-35	1.0
Hydraulic Oils						
Light	160		380	100		
Medium	250		400	100		
Heavy Medium	315		410	100		
Heavy	500		415	95		

Compounded to pass 2500-hr oxidation test D-943, Vickers pump, rust and
foam tests.

TABLE 69 - PROPERTIES OF USED LUBRICATING OIL BEFORE AND
AFTER BUREAU OF MINES TREATMENTS

	Unprocessed Used Crank- Case Oil	Distillation Clay Treatment	Solvent Treatment Clay Clarification	Combined Solvent Treatment Distillation
Elemental Analysis, ppm:				
Ba	312	0	6	0
Ca	2,135	0	212	0
Mg	61	0	53	1
Na	26	0	0	0
Zn	1,994	3	444	10
Pb	12,125	79	83	41
Al	19	0	2	1
Cr	5	0	0	0
Cu	26	5	13	4
Fe	448	0	12	1
Mn	3	0	2	0
Ni	2	1	1	1
Si	49	5	58	27
Sn	11	5	6	0
V	0	0	0	0
Carbon Residue, %	4.68	0.00	0.22	0.00
Sulfated Ash, %	2.08	0.07	0.25	0.05
Nitrogen, %	0.096	0.020	0.060	0.026
Sulfur, %	0.36	0.18	0.21	0.18
Pentane Insoluble, %	2.09	0.02	0.12	0
Antifreeze, ppm	Negative	Negative	Negative	Negative
Saponification Number	16.7	2.96	5.60	3.01
Total Acid Number	6.16	0.38	1.66	0.58
Total Base Number	0.92	0	0.24	0
Gasoline Dilution, %	3.5	0.2	0.6	0.2
Flash Point, °F	370	423	440	427
Pour Point, °F	-35	+15	+5	+15

Operating Plant

Browning-Ferris Industries (BFI) has an oil reclamation
plant at Lemont, Illinois with an annual capacity of over 7 million
gallons per year. Waste crankcase and transformer oils, slop oils
from tank bottoms, tank cleaning, oil spill cleanup, etc. all contribute to the plant feedstock.

The BFI plant can process waste oils with a variety of methods which make the waste oil more economical. These processes can be as simple as heat treatment followed by gravity settling or a combination of heat and chemical treatments followed by centrifuging. The degree of treatment is determined by the original feedstock and the final application.

Figure 54 shows the process flow diagram for the BFI oil recovery plant. The waste oils are received by collection tank trucks and stored in tanks according to type and/or treatment required. Transfer of the oil is by centrifugal pump which, if chemical treatment is necessary, aids in mixing the chemicals and waste oil.

The process is dependent on the input waste oil characteristics. Treatment requirements change with the oil. After the oil passes through a steam heat exchanger and is heated to 180 to 200°F it is tested. If floc formation is great enough to effect gravity

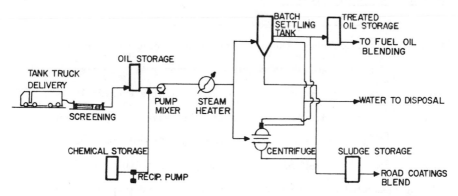

FIG. 54.   The Browning Ferris Industries oil recovery plant.[83]

settling, the oil solution is diverted to the batch settling tank with a cone bottom. If gravity settling is not readily obtainable from the floc formed, the oil solution passes to a centrifuging operation. The purified oil from the centrifuge and the batch settling tanks is collected in storage tanks for eventual delivery to fuel oil users by tank truck.

Wastewaters are collected in an oil-water separator and then trucked to treatment facilities operated by others. The generation of wastewater is low and does not economically warrant an inplant waste disposal system.

Oil sludge from the two processes, the centrifuge and batch treatment tanks, is stored and marketed as a road coating additive. Table 70 shows the typical characteristics of the BFI waste oil input and its treated product. These are compared with typical values for No. 5 and No. 6 fuel oil.[83]

Problems in Waste Oil Recovery

The technology for properly burning or re-refining waste oil is available. However, there are still some by-product or residual product problems which require attention. All the clay treatment processes generate oily solid wastes that are normally land disposed. This material has been tried in several applications including road asphalts, and roofing materials, with minimal success.[84] Studies by Armour Research Institute[85] and Calspan[86] indicate that re-covery of acid sludge may be possible if it can be pretreated, such as water washing it to eliminate water solubles, acid/solvent extrac-tion remove oil, or neutralization with lime to eliminate acidity.

TABLE 70 - COMPARISON OF THE CHARACTERISTICS AND PROPERTIES
OF UNTREATED WASTE OIL, TREATED WASTE OIL AND
COMMON FUEL OILS[83]

	Typical Waste Oil Receipts	Typical Treated Product	#5 Oil Specifications	#6 Oil Specifications
Viscosity, SSV @ 100 F	210-260	200-250	150-300	900-9000
Flash Point, F	150+	150	130 min	150 min
BS&W	4-12	0.4-0.8	1.0 max	2.0 max
Heating Value, BTU/Gal.	*	145,000+		
Pour Point, F	+5	+5		
Sulfur, Wt. %	0.2-0.5	0.2-0.5	1.0 max	1.0 max
Ash, Wt. %	0.3-1.5	0.2-0.8	0.1 max	

*Varies with water content.

Alternatives include pyrolysis to recover the oil fraction and combustion in a cement-type kiln with heat recovery for steam generation.
Both of these will require a neutralization process to eliminate
corrosion. The quality of the recoverable oil from the pyrolysis
process has a similar viscosity to between No. 2 and No. 4 fuel oil
but with a darker color and some odor. The latter combustion concept
is a modification of the Monsanto plant in Baltimore, but a cement
kiln is used in place of the Landguard unit.

        The bottoms residue from the vacuum distillation of
waste oil also requires disposal. Applications as a roofing
material have been acknowledged but leachate in runoff may be
a problem. Utilization of bottoms as a fuel for a secondary
lead reverberatory furnace has been demonstrated on the lab

scale under EPA sponsorship. This disposal method allows for
energy recovery as well as lead recovery. Another method, proposed but not yet demonstrated, would use a bed of molten lead as
a heat source in a pyrolysis unit, recovering fuel values overhead
and trapping impurities (predominantly lead) in the "molten lead
sink."

The most critical problem for recovery and utilization
of waste oil is collection. Studies by the States of Maryland,
New York, Massachusetts, and Wisconsin have determined that adequate
collection is the most inhibiting aspect to properly disposing of
waste oil. Collection from individuals presents an additional
problem over and above those encountered in collecting from source
points such as garages and industrial sites.

Another significant problem, specifically in the utilization of waste lube oils as feedstock for automotive lubricating oils,
is the development of markets which would accept the oil.   The
product must overcome a reputation of poor quality. Automotive
lubricating oil must pass engine use tests in order to comply with
SAE Standards as well as the MIL-Specs. These are costly requirements but are not major obstacles for lubricating oils treated by
a given process from a well defined feedstock. A lube oil generated
from a given blend and based on defined crude sources and processes
requires one set of engine tests. Should these crude sources
change, another engine test is required. This use test has been
used because no other specification test was believed adequate to
judge the lubricating qualities of an oil. These tests are quite

expensive and are seldom performed. Thus, re-refined waste oil

has not been accepted on the automotive lubricating market as being

equivalent to lubricating oils produced from virgin stocks, even if

it was of equally high quality. Alternate methods of specifying

lubricating oils must be defined. Governmental standards are necessary

to guide the industry. A governmental agency such as The Bureau of

Mines, or the SAE should develop and establish physical and chemical

properties and test procedures which could be used to determine quality

and rate reprocessed waste oils. In the interim, waste oil re-refiners

have attempted to show the quality of their product by supporting

projects with municipalities and private industry that use solely

waste lube oil.

Conclusion

Waste oil recovery represents an opportunity for resource

recovery (BTUs, metals,etc.) and a reduction of environmental pollution.

In recent years it has been attractive to merely blend collected waste

oil and burn it as a relatively cheap fuel source. These processes

generate emission problems and EPA regulations will eventually pro-

hibit or at least control the burning of untreated waste oil. Waste

oil combustion, even with heat recovery, does not take into account

the value of the waste oil as a feedstock to produce lubricants.

Use of waste oil as base lubricating stock has a higher market

value than fuels and would replace virgin crudes to be used else-

where. Studies by Sun Oil Company[87] and others show that lubricating

oils are becoming scarce with respect to the rising industrial demand.

Waste oil re-refining may be a solution to these problems provided:

(1) the technologies of the re-refining industry are vastly improved[88]
and (2) more importantly, the governmental action proceeds to give
the re-refiners an economic incentive.[89]

Table 71 is a partial listing of the operating re-refiners
in the United States and Canada.

TABLE 71 - RE-REFINERS - A PARTIAL LIST

Arizona
    Kaibab Industries, Inc., Phoenix 85009

Arkansas
    Henley Oil Company, Norphlet 71759

California
    Bayside Oil Corporation, San Carlos 94070
        Primarily re-refine used automotive crankcase oils using sulfuric
        acid/clay process. Product is recompounded to produce single and
        multigrade compounded and uncompounded motor oils, aircraft engine
        oils, automatic transmission fluids, gear lubricants, and greases.
        Capacity is about 9,000 gallons feed per day. Acid sludge and oily
        clay disposed of in an approved dump.

    Edgington Oil Corporation, Long Beach 90805
        Largest reprocessor of waste automotive oil in California. Process
        about 18,000 gallons of waste oil per calendar day and have capacity
        for twice this amount if waste oil were available. Waste oil is re-
        processed for use as a wide variety of light and heavy fuels and as-
        phalt. No lubricants are produced.

    Fabian Refining Company, Oakland 94601
        Primarily re-refine waste crankcase oils using sulfuric acid/clay
        process. About 11,000 gallons of waste oil is reprocessed per
        calendar day to produce about 7,000 gallons of an SAE 30-grade base
        oil and a small amount of 90 Neutral.

    Leach Oil Company, Inc., Compton 90220
        Re-refine motor oil only using sulfuric acid/clay process. Current
        feed rate is 10,000 gallons per calendar day.

    Motor Guard Lubricants Company, Los Angeles 90023
        Process about 7,000 gallons of waste crankcase oil per calendar day
        using sulfuric acid/clay process.

    Nelco Oil Refining Company, National City 92050
        Re-refine about 4,500 gallons of waste crankcase oil per calendar
        day using sulfuric acid/clay process.

TABLE 71 - (continued)

O and J Refining Company, Fresno 93721
  Reported to have a capacity for re-refining about 10,000 gallons per
  calendar day of product using acid and clay treating. End products
  are used for asphalt, fuel oil, and base lube stock.

Omega Oil Company, Baldwin Park

Talley Brothers, Inc., Huntington Park 90255
  Distillation and clay contacting is used to process about 12,000
  gallons of used oils per calendar day, principally industrial
  oils. Said to be primarily a custom process which returns the
  reclaimed, recompounded oil to its original use.

Colorado
  Custom Refining Company, Commerce City

  Williams Refining Company, Denver 80221
    Has a capacity of about 4,500 gallons per eight hour day. Primari-
    ly re-refines waste crankcase oils.

Connecticut
  Environmental Oil and Liquid Removal, Inc., Prospect

Florida
  Davis Oil Company, Hallandale 32302

  Petroleum Products Company, Hallandale 33009
    Reported to use sulfuric acid/clay process

  Peak Oil Company, Tampa 33619

Georgia
  General Refining Company, Savannah
    Re-refines waste crankcase oils picked up from service stations in
    Charleston, S.C. Present capacity is 9,500 gallons per week
    (reprocessed 360,000 gallons in 1972). Would double capacity if
    waste oil were available in large quantities rather than small
    quantities found at service stations. Disposes of waste products
    in pit on company property.

  Seaboard Industries, Doraville 30340
    Originally used an acid/clay process for re-refining waste crank-
    case oils, but the waste by-products became unacceptable at avail-
    able landfills following a ruling by the State Water Quality Con-
    trol Board. An alkaline process with centrifuge was tried, but the
    reprocessed oil had a dark color that detracts from marketability.

Illinois
  Motor Oils Refining Company, Lyons 60534
    Uses sulfuric acid process to re-refine crankcase and industrial
    oils. Capacity is about 10 million gallons/year.

  Browning-Ferris Industries, Lemont, Illinois
    Uses a variety of processes to re-refine crankcase and industrial
    oils.

TABLE 71 - (continued)

Indiana
    Bates Oil and Refining Company, Chandler
        Small operation using a "re-refined oil machine" that processes
        100 gallon batches of used crankcase oils.

    Smart Oil Company, Environmental Control Div., Indianapolis
        Uses sulfuric acid process mostly for industrial oils. Capacity
        is reported to be about 9 million gallons a year.

    Westville Oil and Manufacturing Company, Westville 46391
        Uses chemical treatment, clay contact and overhead distillation
        to reprocess about 7.5 million gallons annually of used crankcase
        and industrial lubes. Plant has a capacity of 10 million gallons
        and there are plans for a 25 percent expansion. Waste by-products
        are acceptable, after treatment, for disposal in landfills.

Kansas
    Coral Refining Company, Kansas City 66105
        Uses sulfuric acid process to re-refine industrial and crankcase
        oils. Capacity is reported to be about 2.1 million gallons/year.

Michigan
    Dearborn Refining Company, Dearborn 48120
        Uses acid/clay treatment to reprocess industrial and crankcase
        oils. Presently operating at full capacity and expanding. Acid
        sludge is neutralized and disposed of in landfill. Clay also goes
        into landfill.

    Stuart Oil Company, Livonia 48150
        Reprocesses cutting oils by settling, heat and chemical treatment.
        Currently operating at full capacity and considering expansion to
        reprocessing waste crankcase oil.

Minnesota
    Warden Oil Company, Minneapolis 55405
        Uses sulfuric acid and filter system process to re-refine approx-
        imately 600,000 gallons of waste crankcase oils annually. End
        products are mainly for railroad journal boxes, although some motor
        oil is made. Total capacity is 1 million galllons/year.

Mississippi
    Jackson Oil Products, Jackson 39208
        Uses sulfuric acid process to re-refine used motor oils. Plant
        has a capacity of about 3 million gallons/year.

Missouri
    Midwest Oil Refining Company, St. Louis 63114
        Uses thermo dehydration, clay contacting and high temperature dist-
        illation to process all types of waste oils. Products produced in-
        clude: non-additive motor oils, additive type motor oils, equivalents
        to No. 5 heating oils and wood oils. Plant has a capacity of 5 mil-
        lion gallons a year, but presently operating at half capacity due to
        lack of waste oil.

TABLE 71 - (continued)

Montana
    Sears Oil Company, Billings

New Jersey
    Diamond Head Oil Refining Company, Kearny
    Uses caustic process to re-refine both crankcase and industrial
    oils. End products include lubricants and fuel oil. Current
    capacity is approximately 5 million gallons/year.

    National Oil Recovery Corp., Bayonne 07002
    Distillation/hydrotreating process is used to process all types
    of oily wastes. See the paragraphs under "Re-refining" in the
    text for additional details.

New Mexico
    Mesa Oil Company, Inc., Albuquerque
    Used crankcase oil is filtered, light ends flashed off and con-
    ventional sulfuric acid treatment is used to produce re-refined
    motor oil. Plant capacity is 650 gallons/day, but will be ex-
    panded to 1,200 gallons/day.

New York
    Booth Oil Company, Inc., North Tonawanda 14120
    Processes all types of waste oils. Current capacity is approx-
    imately 2 million gallons year.

    Northeast Oil Services Refining Corp., Syracuse

North Carolina
    South Oil Company, Greensboro 27402
    Subsidiary of Seaboard Industries, Doraville, Georgia. Is re-
    ported to use acid/clay process.

Ohio
    Keenan Oil Company, Cincinnati 45212
    Reprocesses all types of waste oils. End products include coal
    spray oil, hydraulic oil and industrial fuel oil but no motor oil.
    Capacity is approximately 8 million gallons/year, but operates
    below capacity because of difficulty in obtaining waste oil.

    Research Oil Refining Company, Cleveland 44109
    Reprocesses industrial oils only. No process information available.

Oklahoma
    Double Eagle Refining Company, Oklahoma City 73111
    Conventional sulfuric acid process. Handles mostly motor oils and
    a small volume of industrial oils. Operating below capacity of
    approximately 5-6 million gallons a year.

Oregon
    Ager and Davis Refining Company, Portland 97211
    Uses sulfuric acid process to re-refine mostly crankcase oils,
    but some industrial oils. Capacity is estimated at 1.8 million
    gallons a year.

    Nu-Way Oil Company, Portland 97218

TABLE 71 - (continued)

Pennsylvania
   Berks Associates, Pottstown 19464
      Uses caustic-naphtha process. Re-refines mostly motor oil, but
      some industrial oil. Would like more waste oil for reprocessing
      and would be willing to expand if the waste oil were available.
      Capacity is about 12.8 million gallons a year, but operates at
      approximately 65 percent of capacity.

   Petrocon Corporation, Modena, Chester County
      No information available on process used, but it does not involve
      acid treatment. Processes industrial and automotive waste oils
      to produce fuel oil and some lubricants. Capacity is in excess
      of 15 thousand gallons per 24 hour day. Indicates no problems
      with waste by-products.

Tennessee
   Gurley Oil Company, Memphis 38102
      Re-refines all types of waste oils using conventional sulfuric
      acid process to produce lubricating oils. Capacity is 5-6 million
      gallons/year. Waste by-products are disposed of in sanitary land-
      fill.

Texas
   Capital Supply Company, Hurst 76053
      Mainly processes waste crankcase oils to produce automotive lubes.
      Plant capacity is approximately 150,000 gallons per month. Process
      includes various methods such as filter clays, etc., but no acid
      use reported.

   Cooks Oil Company, Boyd
      Has a plant capacity of about 60,000 gallons per month. Waste
      crankcase oil is primary feed stock. End products include trans-
      mission and hydraulic fluids. No information available on process
      used.

   Texas American Petrochemicals Inc., Midlothian
      Reprocesses mainly waste crankcase oils to produce base stock that
      is sold in bulk to jobbers. No blending is done to achieve a
      given SAE grade. Capacity is approximately 250,000 gallons per
      month. No information available on process used.

   S&R Oil Company, Houston 77035
      Uses several techniques -- dehydration, acid treatment, clay
      treatment, distillation -- to produce end products that consist
      mainly of automotive lubricants. Uses any lube based material
      as a feed stock, but mainly waste crankcase oils. Present cap-
      acity is about 600 thousand gallons per month, but processes
      about 300 thousand gallons a month based on a 12 hour day, 5 day
      week operation. Waste by-products are dumped in landfills or
      mixed with aggregates for use on roads.

TABLE 71 - (continued)

Utah
    Alco Refining Division of Bonus International Corp., Salt Lake City,
    84116
        Sulfuric acid process is used to re-refine used motor and indus-
        trial oils. End products include both automotive lubes and in-
        dustrial oils. Capacity is approximately 5,500 gallons per day.
        Waste products are disposed of in Salt Lake City landfill dump.

Washington
    Superior Refineries, Inc., Woodinville 98072

    OED Corporation, Renton, 98055

Wisconsin
    Warden Refining Company, West Allis 53214
        Reprocesses used industrial and motor oils by hydration, chemical
        treatment, distillation and filtration. End products are motor,
        road and fuel oils. Total capacity is 1,000 gallons per day.
        Currently processes about 2,500-3,000 gallons per week.

Canada
    Hub Oil Company, Ltd., SE Calgary Canada T2A OW4
        Primarily re-refine used automotive crankcase oils, using sulfuric
        acid/clay process. Products recompounded to produce motor oils,
        gear oils, hydraulic oils, etc. Operating below yearly capacity of
        3 million gallons.

Pyrolysis has recently become a viable alternative as a process for resource recovery of solid wastes. There are many advantages of the pyrolysis process over incineration including:

1) Pyrolysis plants require a smaller capital investment and lesser operating costs because of their lower operating temperatures. Materials of construction and maintenance are less costly on a pound for pound basis.

2) Pyrolysis plants currently in operation for other processes are 5 to 10 times larger than conventionally sized incinerators. Quantities of solid wastes by nearly an order of magnitude larger than normal can be processed in a pyrolysis plant which requires a relatively smaller investment.

3) The large air transport systems of incinerators are not required in a pyrolysis plant.

4) Thermal efficiencies obtained at a pyrolysis plant are greater than those obtained at incinerators.

5) Air pollution devices may not be necessary at pyrolysis plants and therefore, air pollution control costs that can be 25 percent of some plant's initial costs can be reduced.[50]

Pyrolysis techniques include the capability of operating under conditions of no oxygen or partial oxygen or combustion. The pyrolysis of municipal solid wastes has been considered in fluid bed reactors or fixed bed reactors. Figure 55 shows a pyrolysis system utilizing a fluid bed reactor for handling solid wastes. Fluidized reactor beds can offer advantages for processing municipal solid wastes because of the wastes' general heterogeneous composition (i.e., aluminum and glass). As a result of

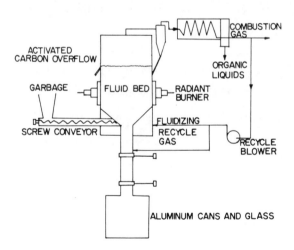

FIG. 55. Pyrolysis system for handling municipal garbage.[9]

velocity differentials among the various components within the
pyrolytic reactor, the bed movement acts as a waste classifier
as well as energy source. Glass particles and aluminum cans
settle to the reactor bottom and are removed, organics rise to the
top of the reactor unit and are processed.

A pyrolysis system developed for solid waste disposal
involves a two-stage unit. Sand circulates between the two stages:
a combustion stage which generates heat to effect the pyrolysis
stage where solid wastes are decomposed. This process produces an
activated char which can be combusted in the first stage and used
to treat the various by-products of the system. Gases effected by
the final pyrolytic action can be directly combusted or treated
for pollutant removals before utilization.

Pyrolysis has had applications in the petroleum and
paper and pulp industries. It has also been researched and developed for coal gasification processing which has eventually
lead to present solid waste disposal and recovery methods.[90]

Pyrolysis Process

Pyrolysis is an endothermic process which heats
organic materials to high temperatures without oxygen resulting
in the breakdown of these organic materials into their various
components. Pyrolysis temperatures are usually in the range
between 500 to 1100°C. These temperatures, no oxygen and most
organic materials effect the formation of three product types:

1)   a gas consisting of hydrogen, methane, carbon
monoxide and carbon dioxide.

2)   a liquid mixture consisting of water, tar, oil and
other organics.

3)   a solid residue consisting of carbon and ash.

The products of pyrolysis can be used as a fuel, except
for the water and ash, and with minimal pollution effects. Other
products such as acetic acid, methanol and other solvents can be
resold.

The Bureau of Mines[90] has extensively studied the conversion of municipal and industrial solid wastes into useful
materials. They experimented with four different feedstocks.

1)   Raw municipal wastes, with metals and glass removed,
was shredded and used.

2)   Plastic materials were removed from the raw wastes by
an air classification and screening process and used (with plastic
film).

3)   Industrial wastes with metals removed were shredded
with two different types of equipment and as such were tested
separately.

a)   Heil mill shredded industrial wastes.

b)   Gondard mill shredded industrial wastes.

An analysis of each of these feedstocks before pyrolysis is shown
in Table 72. Each feedstock was pyrolyzed in the state noted above.
Note the differences in moisture and ash contents of municipal solid
wastes as compared to industrial wastes. Moisture content of

TABLE 72 - AVERAGE ANALYSES OF REFUSE USED IN PYROLYSIS TESTS[90]

	Raw Municipal Refuse		Processed Municipal Refuse Containing Plastic Film		Heil Mill Industrial Refuse		Gondard Mill Industrial Refuse	
	As Received	Dry	As Received	Dry	As Received	Dry	As Received	Dry
Proximate, Percent:								
Moisture	43.3		42.3		20.5		33.2	
Volatile Matter	43.0	76.3	44.3	76.8	40.3	49.7	33.5	50.6
Fixed Carbon	6.7	11.7	5.6	9.7	9.9	12.3	4.6	7.0
Ash	7.0	12.0	7.8	13.5	29.3	38.0	28.7	42.4
Total	100.0	100.0	100.0	100.0	100.0	100.0	100.0	100.0
Ultimate, Percent:								
Hydrogen	8.2	6.0	7.6	5.0	6.0	4.6	6.2	3.8
Carbon	27.2	47.6	27.2	47.3	27.5	33.9	20.6	31.2
Nitrogen	.7	1.2	.8	1.4	.5	.7	.5	.7
Oxygen	56.8	32.9	56.5	32.6	36.4	22.4	43.9	21.8
Sulfur	.1	.3	.1	.2	.3	.4	.1	.1
Ash	7.0	12.0	7.8	13.5	29.3	38.0	28.7	42.4
Total	100.0	100.0	100.0	100.0	100.0	100.0	100.0	100.0
BTU per Pound of Refuse	4,827	8,546	5,310	9,180	4,570	5,645	3,415	5,155
Available BTU Per Ton of Refuse, Millions	9.654	17.092	10.620	18.360	9.140	11.290	6.830	10.310

271

municipal wastes was about 43 percent and ash content about 7 to 8

percent. Industrial moisture content ranged between 20 to 30 per-

cent with a 29 percent ash content. Heating values varied from

9.6 to 10.6 million BTU per ton for municipal solid wastes and

from 6.8 to 9.1 million BTU per ton for industrial wastes.

Figure 56 shows the flow diagram of the Bureau of Mines pilot

plant used to pyrolyze the wastes. An electric furnace with

nickel-chromium resistors accepts the refuse, is sealed, and then

heated. Gases and other vapors generated by the process pass into

an air cooled trap where heavy oils and tar are condensed and re-

moved. Two water-cooled condensors cool the remaining gases and

vapors to room temperature for removal of additional heavy oil and

liquor. Heavy oil mists are removed by the electrostatic pre-

cipitators and the gases then pass through a series of packed

scrubbers for ammonia, carbon dioxide and hydrogen sulfide removal.

The gases are analyzed and flared (under normal operations this

gas would be utilized in furnaces, boilers, etc.).

Pyrolytic tests, depending on waste type and temperature,

ranged from 6 to 12 hours in length. The initial test temperature

was 500°C which was increased to 2 to 3°C per minute up to 900°C.

Other tests were run at constant temperatures of 750°C and 900°C.

Pyrolytic Products - All the products of the pyrolytic

action were collected -- tar, heavy oils, light oils, liquors

and other residues -- and analyzed. Table 73 gives the various

products obtained from pyrolysis. At 900°C 154 pounds of solid

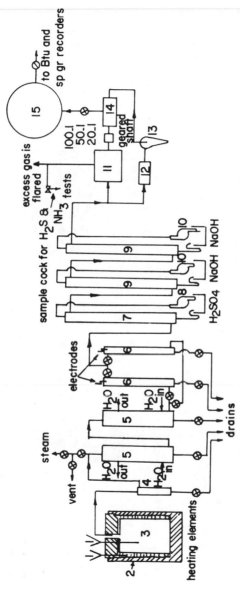

FIG. 56. Flow diagram of U. S. Bureau of Mines pyrolysis system. 90 1, thermocouple; 2, electric furnace; 3, retort; 4, tar trap; 5, tubular condenser; 6, electrostatic precipitator; 7, ammonia scrubber; 8, acid pump; 9, carbon dioxide and H₂S scrubber; 10, caustic pump; 11, large wet-test meter; 12, drying tube; 13, light-oil condenser; 14, small wet-test meter; 15, gas sample holder.

TABLE 73 - YIELDS OF PRODUCTS FROM PYROLYSIS OF MUNICIPAL AND INDUSTRIAL REFUSE[90]

Refuse	Pyrolysis Temp., °C	Yields, Weight-Percent of Refuse							Yields Per Ton of Refuse				
		Residue	Gas	Tar	Light Oil In Gas	Free Ammonia In Gas	Liquor	Total	Gas, Cubic Feet	Tar, Gallons	Light Oil In Gas, Gallons	Liquor, Gallons	Ammonium Sulfate, Pounds
Raw Municipal	500-900	9.3	26.7	2.2	0.5	0.05	55.8	94.6	11509	4.8	1.5	133.4	17.9
	750	11.5	23.7	1.2	.9	.03	55.0	92.3	9628	2.6	2.5	131.6	23.7
	900	7.7	39.5	.2	-	.03	47.8	95.2	17741	.5	-	113.9	25.1
Processed Municipal Containing Plastic Film	500-900	21.2	27.7	2.3	1.3	.05	40.6	93.2	11545	5.6	3.7	96.7	16.2
	750	19.5	18.3	1.0	.9	.02	51.5	91.2	7380	2.2	2.6	122.6	28.4
	900	19.1	40.1	.6	.2	.04	35.3	95.3	18058	1.4	.6	97.4	31.5
Heil Mill Industrial	500-900	36.1	23.7	1.9	.5	.05	31.6	93.9	9563	4.1	1.4	75.2	12.5
	750	37.5	22.8	.7	.9	.03	30.6	92.5	9760	1.5	2.6	73.0	19.5
	900	38.8	29.4	.2	.6	.04	21.8	90.8	12318	.5	1.6	51.1	21.7
Gondard Mill Industrial	500-900	41.9	21.8	.8	.6	.03	29.5	94.6	9270	1.7	1.6	70.2	20.4
	750	31.4	25.5	.8	.8	.03	31.5	90.0	10952	1.8	2.2	74.9	21.2
	900	30.9	31.5	.1	.5	.03	29.0	92.0	14065	.02	1.4	68.5	22.9

residue, one-half gallon of tar, 114 gallons of liquor, 25 pounds
of ammonium sulfate, and 17,741 cubic feet of gas were the products
from pyrolyzing one ton of raw municipal solid wastes. At 750°,
the gas generated by the pyrolysis process was decreased by almost
50 percent, but tar and light oil production increased dramatically.
Pyrolysis with a temperature gradient from 500°C to 900°C effected
the largest quantities of tar, but a gas yield of 11,509 cubic feet.

Municipal refuse with plastic film pyrolyzed at 900°C
resulted in 382 pounds of solid residue, 1.4 gallons of tar, 0.6
gallons of light oil, 97.4 gallons of liquor, 31.5 pounds of ammonium sulfate, and 18,058 cubic feet of gas. Again, at 750°C the gas
evolved decreased by almost 50 percent with tar and light oil quantities increasing. Pyrolysis through a temperature gradient (500 -
900°C) of the wastes with plastics yielded relatively similar results as raw refuse.

Industrial wastes comprised mainly of paper products,
rags, and small quantities of metals yield much greater amounts
of solid residue (618 pounds), yet comparable amounts of the other
products, per ton of waste processed. Pyrolysis temperature differences did not markedly change the quantities of gas obtainable
from industrial wastes.

Solid Residue - The solid residue analysis is given in
Table 74. The pyrolization of municipal solid wastes which was
processed and contained plastics resulted in a char residue with a
fixed carbon content of nearly 57 percent and a heating value of
nearly 18 million BTU per ton. The industrial wastes produced

TABLE 74 - CHEMICAL ANALYSES[1] OF SOLID RESIDUES FROM PYROLYSIS OF MUNICIPAL AND INDUSTRIAL REFUSE[90]

Refuse	Pyrolysis Temp., °C	Proximate, Percent				Ultimate, Percent					Heating Value, BTU/Lb	Heating Value, Million BTU/Ton
		Moisture	Volatile Matter	Fixed Carbon	Ash	Hydrogen	Carbon	Nitrogen	Oxygen	Sulfur		
Raw Municipal	500-900	2.6	4.4	29.6	66.0	0.4	32.4	0.5	0.5	0.2	5,020	10.040
	750	2.2	7.4	51.4	41.2	.8	54.9	1.1	1.8	.2	8,020	16.040
	900	1.0	4.7	31.7	63.6	.3	36.1	.5	.0	.2	5,260	10.520
Processed Municipal Containing Plastic Film	500-900	1.7	4.8	56.7	38.5	.6	57.7	.8	2.1	.3	8,800	17.700
	750	1.3	13.4	34.6	52.0	.8	41.9	.8	4.4	.1	6,080	12.160
	900	1.2	3.3	53.5	43.2	.5	53.4	.7	1.8	.4	8,090	16.180
Heil Mill Industrial	500-900	.9	2.6	15.2	82.2	.3	17.0	.1	.2	.2	2,520	5.040
	750	1.2	5.1	17.0	77.9	.5	19.4	.2	1.8	.2	2,900	5.800
	900	.1	2.5	12.9	84.6	.3	14.8	.2	.0	.2	2,180	4.360
Gondard Mill Industrial	500-900	.3	3.0	9.7	87.3	.2	11.8	.1	.4	.2	1,660	3.320
	750	1.0	3.6	16.6	79.8	.3	19.5	.2	.0	.2	2,680	5.360
	900	.2	6.4	16.2	77.4	.4	19.3	.3	2.4	.2	2,810	5.620

[1]Moisture on as-received basis; all other data on dry basis.

chars low in both fixed carbon (9 to 17 percent) and heating value
(3.3 to 5.8 million BTU per ton).

Not only can energy be recovered from the products of
pyrolysis, but the volume of wastes is dramatically reduced.
Municipal solid wastes are reduced by 90 percent, municipal solid
wastes with plastics are reduced by 80 percent, and industrial
solid wastes are reduced by 65 percent their original volume.

Gases - Table 75 gives the analysis of the gases evolved
from the pyrolysis process. Hydrogen, carbon monoxide, methane,
ethylene and carbon dioxide are the major components of the gas.
Gas generated by pyrolysis at 750°C had the highest heating values
for all types of wastes. However, the larger quantities of gas
generated by pyrolysis at 900°C resulted in larger amounts of heat
being available per ton of solid waste processed. Additional
treatment of the gases may result in higher heating values.

Pyrolyzation of one ton of municipal solid waste at
900°C requires 2 million BTU and this process generates gases
with about 8 million BTU available for utilization. Therefore,
the process is not only self-sustaining, but produces a fuel for
other uses.

Tar - Pyrolysis at 900°C of the wastes resulted in
neglible amounts of tar. The analysis of the tar generated is
given in Table 76. Depending on the temperature of pyrolysis and
the type of waste pyrolyzed, the content of neutral oil in the
tar produced varied from below 20 percent for raw municipal wastes

TABLE 75 - ANALYSES OF GASES FROM PYROLYSIS OF MUNICIPAL AND INDUSTRIAL REFUSE[90]

	Raw Municipal Refuse			Processed Municipal Refuse Containing Plastic Film			Heil Mill Industrial Refuse			Gondard Mill Industrial Refuse		
Pyrolysis Temp., °C	500-900	750	900	500-900	750	900	500-900	750	900	500-900	750	900
Analysis, Vol. Percent:												
Hydrogen	45.47	30.86	51.91	44.86	25.27	42.41	45.50	47.89	49.12	47.45	49.63	51.14
Carbon Monoxide	21.54	15.57	18.16	19.62	25.09	20.16	20.67	13.04	19.39	20.70	12.14	18.19
Methane	13.15	22.57	12.66	18.73	17.57	13.92	16.68	20.27	15.94	12.59	17.09	12.61
Ethane	1.30	2.05	0.14	2.08	2.01	0.25	1.27	1.50	0.37	1.49	1.25	0.32
Ethylene	4.67	7.56	4.68	4.54	10.36	7.89	2.77	3.97	3.86	3.39	5.30	4.42
Carbon Dioxide	11.41	18.44	11.42	8.02	18.25	13.91	10.51	11.99	10.25	10.24	12.93	11.71
Propane	<0.01	<0.01	<0.01	<0.01	<0.01	1.17	<0.01	<0.01	<0.01	<0.01	<0.01	<0.01
Propylene	1.32	1.53	0.32	1.35	0.76	0.10	1.38	0.86	0.34	2.35	0.89	0.62
Isobutane	Trace	Trace	Trace	Trace	Trace	Trace	Trace	Trace	Trace	Trace	Trace	Trace
Butane	0.08	0.01	0.44	0.03	Trace	0.11	0.05	0.01	<0.01	0.13	0.02	<0.01
Butene-1	0.16	0.15	Trace	0.11	Trace	Trace	0.15	0.04	Trace	0.16	0.05	0.17
Isobutylene	0.17	0.15	Trace	0.15	Trace	Trace	0.23	0.07	Trace	0.25	0.12	Trace
trans-Butene-2	0.04	0.03	Trace	0.08	Trace	Trace	0.12	0.02	Trace	0.02	0.02	0.01
cis-Butene-2	0.07	Trace	Trace	0.03	Trace	Trace	0.07	0.01	Trace	0.01	0.01	0.01
Pentane	Trace	Trace	Trace	Trace	Trace	Trace	Trace	Trace	Trace	Trace	Trace	Trace
Pentenes	0.60	0.87	0.20	0.17	0.53	0.07	0.55	0.29	0.63	0.70	0.51	0.39
Unidentified	<0.01	0.21	0.06	0.06	0.15	0.01	0.04	0.03	0.08	<0.01	0.03	0.36
BTU/Cubic Foot of Gas	473	563	447	536	570	511	478	518	498	471	502	447
Million BTU/Ton of Refuse Pyrolyzed	5.473	5.421	7.930	6.188	4.207	9.228	4.571	5.056	6.134	4.366	5.498	6.067

Note: <0.01 = 1 part in 10^4. Trace = less than 1 part in 10^5.

TABLE 76 - ANALYSES OF TARS FROM PYROLYSIS OF MUNICIPAL AND INDUSTRIAL REFUSE BY COAL-TAR METHOD[90]

	Specific Gravity at 15.6°C / 15.6°C	Weight-Percent of dry tar		Boiling Range, Volume Percent						Distillate, Volume - % of Dry Tar			Neutral Tar Oil Volume - %		
		Anthracene	Naphthalene	0° To 170°C	170° To 235°C	235° To 270°C	270° To 350°C	To 500°C	Resi-due 500°-900°C	Acids	Bases	Neutral Oil	Ole-fins	Aro-matics	Paraffins and Naphthenes
PYROLYSIS TEMPERATURE, 500° - 900° C															
Raw Municipal	1.077	0.0	0.0	4.4	12.4	7.2	20.0		56.0	7.4	4.7	31.9	17.5	68.1	14.4
Processed municipal containing plastic film	.974	.0	.0	1.7	9.1	7.0	19.3		62.9	5.2	3.7	28.0	25.0	62.6	12.4
Heil Mill Industrial	1.111	.0	.0	3.3	11.1	5.4	10.2		70.0	5.9	3.7	20.4	23.0	68.5	8.5
Gondard Mill Industrial	1.093	.0	.0	3.7	14.1	7.4	16.5		58.3	7.0	5.3	29.1	21.3	67.7	11.0
PYROLYSIS TEMPERATURE, 750° C															
Raw Municipal	1.115	0.59	3.17	1.9	11.1	4.1	8.1		74.8	4.0	2.1	17.5	16.3	79.9	3.8
Processed municipal containing plastic film	1.101	Trace	Trace	4.2	13.2	7.5	12.0		63.1	4.9	6.3	25.7	23.5	69.6	6.9
Gondard Mill Industrial	1.099	.73	4.07	6.2	26.8	8.4	14.3		44.3	6.8	6.0	38.3	19.0	78.2	2.8

TABLE 77 - YIELDS OF TAR COMPONENTS FROM PYROLYSIS OF MUNICIPAL AND INDUSTRIAL REFUSE[90]

Refuse	Gallons Per Ton of Refuse								Pounds Per Ton of Refuse	
	Tar	Acids	Bases	Neutral Oil	Residue	Olefins	Aromatics	Paraffins and Naphthenes	Anthracenes	Naphthalenes
	PYROLYSIS TEMPERATURE, 500° - 900° C									
Raw municipal	4.8	0.4	0.2	1.5	2.7	0.3	1.1	0.2	-	-
Processed municipal containing plastic film	5.6	.3	.2	1.6	3.5	.4	1.0	.2	-	-
Heil Mill Industrial	4.1	.2	.2	.8	2.9	.2	.6	.1	-	-
Gondard Mill Industrial	1.7	.1	.1	.5	1.0	.1	.3	.1	-	-
	PYROLYSIS TEMPERATURE, 750° C									
Raw municipal	2.6	0.1	0.1	0.5	2.0	0.1	0.4	0.02	0.14	0.76
Processed municipal containing plastic film	2.2	.1	.1	.6	1.4	.1	.4	.04	Trace	Trace
Gondard Mill Industrial	1.8	.12	.1	.7	.8	.1	.5	.02	.12	.67

TABLE 78 - ULTIMATE ANALYSIS OF TARS FROM PYROLYSIS OF MUNICIPAL AND INDUSTRIAL REFUSE, DRY BASIS[90]

Pyrolysis Temp. °C	Raw Material Refuse			Processed Municipal Refuse Containing Plastic Film			Heil Mill Industrial Refuse			Gondard Mill Industrial Refuse		
	500-900	750	900	500-900	750	900	500-900	750	900	500-900	750	900
Analysis, Percent:												
Hydrogen	8.2	6.9	7.3	10.3	9.3	8.9	7.6	7.0	7.8	8.3	7.0	8.0
Carbon	80.9	84.7	85.0	83.0	86.1	86.9	79.7	83.3	55.0	82.6	85.2	59.7
Nitrogen	2.5	2.8	2.4	1.2	1.1	1.3	2.8	2.7	2.1	2.6	2.4	2.2
Oxygen	6.9	3.8	2.3	5.1	3.2	2.6	7.0	5.1	31.6	4.9	4.2	27.9
Sulfur	0.4	0.5	0.7	0.1	0.3	0.2	0.6	0.7	0.5	0.3	0.4	0.5
Ash	1.1	1.3	2.3	0.3	0.0	0.1	2.3	1.2	3.0	1.3	0.8	1.7
BTU/Pound of Tar	15870	15800	16005	17590	17630	17250	15350	15140	10160	16150	16050	11270

at 750°C to nearly 40 percent for the same at temperatures increasing from 500°C to 900°C. The amounts of each tar component generated
per ton of solid waste pyrolyzed are given in Table 77. The yield
of tar components can vary drastically with temperature and the
nature of the waste. However, as seen in the ultimate analyses
of the tars in Table 78, the relative amounts of the elemental
constituents are about the same for all operating conditions and
wastes. Tons of municipal waste origin generally have higher heating
values than the industrial waste tars.

Light Oils and Liquor

Light oils generated from the pyrolysis process were
primarily benzene. The amounts of benzene available were dependent on the heating process. Increases in benzene content of over
100 percent were observed for raw municipal solid wastes first
allowed to heat gradually from 500°C to 900°C and then heated at a
constant temperature of 750°C. In all cases, the gradual heating
of the waste produced a light oil with a benzene content significantly lower than that produced at constant temperatures.
Table 79 shows the light oil analysis from the pyrolysis of
municipal and industrial solid wastes.

Liquor collected in the water condensers and tar traps
was primarily water. Municipal solid waste pyrolysis effected a
liquor with a water content of 94 to 100 percent. Industrial
solid waste pyrolysis effected a liquor which was about 90 to 97
percent water. Table 80 shows the water content of the liquor
generated by the pyrolysis processes tested.

TABLE 79 - CHROMATOGRAPHIC ANALYSES OF LIGHT OILS FROM PYROLYSIS OF MUNICIPAL AND INDUSTRIAL REFUSE, VOLUME PERCENT[90]

Pyrolysis Temp. °C	Raw Municipal Refuse			Processed Municipal Containing Plastic Film			Heil Mill Industrial Refuse			Gondard Mill Industrial Refuse		
	500-900	750	900	500-900	750	900	500-900	750	900	500-900	750	900
Pre-Benzene	25.04	3.70	0.80	28.08	0.74	0.79	14.25	11.05	0.45	22.88	5.33	3.38
Benzene	37.54	78.47	73.39	57.35	85.18	92.10	60.89	70.63	88.29	51.07	76.25	89.00
Toluene	23.76	14.06	12.25	10.31	11.54	3.82	15.96	14.19	6.40	15.93	14.16	5.10
Ethylbenzene	2.50	.31	.02	1.38	.14	<.01	2.23	.42	<.01	2.38	.30	<.01
m,-p, Xylene	2.39	.65	2.84	.71	.47	.57	1.43	.87	.86	1.42	1.09	.57
O-Xylene	1.01	.20	.81	.36	.20	.16	.59	.40	.12	.68	.35	.10
Unidentified	7.76	2.61	9.89	1.81	1.73	2.45	4.65	2.44	2.88	5.64	2.52	1.84

TABLE 80 - WATER CONTENT OF LIQUOR PRODUCED IN THE PYROLYSIS
OF MUNICIPAL AND INDUSTRIAL REFUSE[90]

| | | Percent water in-- | |
Refuse	Pyrolysis Temp., °C	Liquor From Condensers	Liquor From Tar Trap
Raw Municipal	500-900	96.1	96.2
	750	98.2	94.5
	900	98.9	94.6
Processed Municipal Containing Plastic Film	500-900	97.4	99.7
	750	100.0	98.9
	900	99.3	91.5
Heil Mill Industrial	500-900	91.4	94.5
	750	90.1	95.1
	900	96.8	92.0
Gondard Mill Industrial	500-900	95.0	95.2
	750	94.2	95.4
	900	92.2	94.5

Although the final products of any pyrolysis process would be dependent on operating conditions and feedstock characteristics, several comments can be made concerning its utilization.

1) Pyrolytic processing of municipal and selected industrial wastes is feasible.

2) The process is not only self-sustaining but evolves products with available energy contents.

3) Sulfur content of these products is relatively low (see Tables 74 and 78).

4) Sanitary landfill space requirements can be significantly reduced.

5) Minimum processing for materials with densities of
15-20 lb/ft^3 would occur at temperatures around 800°C.

6) Over twenty compounds were identified as available
in pyrolyis products (see Tables 72 through 80).

7) The fixed carbon residue (char) can be treated into
activated charcoal with exceptional adsorption properties.

8) Plastics usually are converted into carbon and gas
at about 800°C.

9) At 800°C glass begins to melt but does not turn
fluid.   (Aluminum cans are affected in a similar manner.)

10) Magnetic metals are unaffected.[90,91]

The utilization of pyrolysis as a disposal method and
the ultimate energy recovery of municipal solid wastes is dependent
on many criteria.

1) The cost of any operation is of particular importance
and is the major limiting factor of most municipal projects.
Assuming the cheapest landfill operation requires 1 unit cost per
ton of solid waste processed, Table 81 shows the relative costs of
the various disposal methods.   Costs of sanitary landfilling are
relatively inexpensive.   However, landfill is rapidly becoming
scarce, and more and more wastes are being landfilled that were
previously ocean dumped (i.e. sewage sludge, industrial sludge),
resulting in continually rising costs.   Composting is both econo-

TABLE 81 - RELATIVE COST OF SOLID WASTE DISPOSAL FOR
VARIOUS DISPOSAL METHODS

Method	Relative Cost Per Ton Solid Waste
Sanitary Landfill	1.0 - 3.3
Incineration	5.3 -10.7
Composting	6.7 -13.3
Pyrolysis	5.3 -10.7

mically and technically unfeasible for large scale waste disposal.
Incineration and pyrolysis effect similar disposal costs. However,
incineration requires many pollution control devices to meet
governmental standards.   As more stringent regulations are promulgated, incineration pollution control costs will rise proportionately
to the degree of control required. Pyrolysis is inherently
pollution free and produces diversified products for many
applications making their sale more likely and generating revenues to lower operating costs.

2) Volume reduction is a critical problem not only
because of rising landfill costs, but because there is a scarcity
of available landfill sites. Usual compaction for volume reduction prior to landfilling is easily surpassed by incineration and
pyrolysis.   Pyrolysis usually effects a 10 percent greater volume
reduction than incineration thus requiring a lesser transportation
cost than any method.

3) Environmental regulations are increasingly diminishing landfill sites and incinerators. Engineering of a sanitary landfill is becoming more and more expensive as it becomes more difficult to meet regulations. Costs for engineering a landfill for pyrolysis char are quite low because of its stability. Available pyrolyzing units are based on a concept of extraction, trapping, utilization and/or careful disposal of effluents and the stabilization of other products.

4) Investments in sanitary landfills and incinerators are fairly stable, but their usage is limited by significant operational disadvantages. Pyrolysis projects are not only stable investments but produce a marketable product. The attractiveness of a pyrolysis system is exemplified by the following:

. Low dependence on external fuel sources to sustain the process.

. Disposal of many hazardous wastes.

. Conversion of wastes into energy sources.

. Marketability of fuel products.

. Environmentally safe process.

The major problems with pyrolysis is its usual requirements of the solid waste to pass through shredding and classification processes prior to heat treatment. Revenues per pound of fuel processed can be substantially increased if the proper sorting techniques are employed. Magnetic metals are easily removed, but the extraction of other metals and glass usually is a problem and is desirable for high efficiency pyrolysis.

The fluidized bed has been developed for use with pyrolysis. One such fluidized bed method for simultaneous waste classification and pyrolysis was previously described and illustrated in Figure 55. Figure 57 is a schematic diagram of another fluidized bed for pyrolysis method. Waste particles are first ground to a uniform size and injected into the fluid bed of sand. The sand is kept fluid by the flow of hot gases through it, which also is used as an air classifier for those materials not pyrolyzed. Materials are sorted by the gas stream and withdrawn from different levels of the reactor. The separation and grinding processes may prove to be economically unfeasible and the sand bed sometimes fuses at high temperatures. Pyrolysis technology is not complete as research for an optimum bed size, type, and temperature for a solid waste of optimum size and nature.

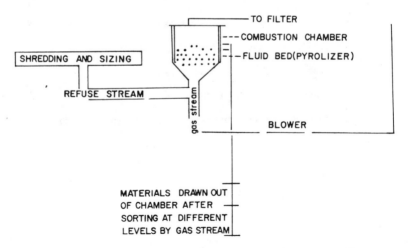

FIG. 57. Schematic of fluid bed pyrolysis process flow.

Pyrolysis Processes

Five major pyrolysis processes devoted to the disposal of municipal refuse and energy recovery are now available. A few of these units were previously discussed and each unit has at least been demonstrated on the pilot plant scale. These processes include:

1) Lantz Converter

2) Union Carbide Refuse Converter

3) Battelle Pyrolizer

4) Garrett Pyrolysis Process

5) Monsanto Landgard System

The Lantz Converter

Developer:	Pan American Resources, Inc. Albuquerque, N.M.
Status:	50 Ton/Day plant in operation at Ford Motor Co., San Jose, California
Recovery:	Gas is recycled to sustain the process, char
Capacity:	1/4-12 tons solid waste per hour available; 5000 Ton/Day plant proposed
Economics:	Based on a 5000 Ton/Day solid waste plant Capital Costs: $32.6 Million Operating Costs: Not available Revenues: Not available

The Lantz converter process consists of a shredding operation in which the solid waste is ground up and then passes into an inclined stainless steel drum. Natural gas is used to prime the process heating it to about 650°C. The gases evolving from the pyrolysis process are fed back to the gas burners to sustain the process.

Only 70 percent of the off gas is required to make the process
self-sustaining. Presently, the remaining gas is scrubbed and
flared. Should a local market become available, this flared gas
will be recovered and sold.

The Union Carbide Refuse Converter

Developer:	Union Carbide Linde Division Tarrytown, New York
Status:	5 Ton/Day solid waste pilot plant in operation
Recovery:	Gas to sustain the process and available as a fuel
Capacity:	150 Ton/Day solid waste demonstration plant proposed in Mt. Vernon, N.Y.
Economics:	Capital investment and operating cost estimates are: substantially lower than those for an incinerator of equivalent capacity. Based on a 1000 ton solid waste per day capacity, the net cost estimates are $1.3 million/yr.

Solid waste is continuously fed into a combustion chamber with a
partial atmosphere. Gases evolving from the process are passed
through an electrostatic precipitator to effect oil and ash removal which are returned to the combustion chamber. The treated
gas stream is either utilized to sustain the process or is neutralized in a caustic solution. The final gas emitted contains particulate matter in the order of 10 percent of that permitted by 1971
Federal Standards, but, its heating value is only 1/2 that of natural
gas. Operating temperatures may reach 1400°C melting most non-combustibles. The slag is drawn off and can be separated into its
metallic and ceramic components for product recovery.

The Battelle Pyrolizer

Developer:	Battelle/EPA Sponsored Pacific N.W. Laboratories Richland, Washington
Status:	Pilot plant to process 10 tons solid waste per day in operation
Recovery:	Fuel gas
Capacity:	100-200 tons of solid waste per day proposed demonstration
Economics:	Not available

The basic pyrolysis process of the pilot plant utilizes a verticle combustion chamber with solid wastes entering the top. The solid wastes are fed through the chamber by gravity with the upper fraction representing the chamber drying zone, the middle fraction the chamber charring zone and the lower fraction the chamber combustion zone. Gases evolving from the process pass up through the solid wastes initiating the process. The gases are then treated and flared.

The Garrett Pyrolysis Process

Developer:	Occidental Research Corporation (Garrett Research & Development Co.) La Verne, California
Status:	Demonstration plant under construction in San Diego County, California
Recovery:	Oil, gas, char and ferrous metals
Capacity:	Input-200 tons solid wastes per day (demonstration)
	Output-Based on 2000 tons solid waste per day proposal

Oil -   480 T/D
Char -   160 T/D
Glass -   120 T/D
Metals -   140 T/D

Economics: Based on 1972 cost estimates:

                          Basis: 2000 T/D - 700,000 Tons/Year

                          Capital Costs: $13,920,000.

                          Operating Costs: $2,701,000/yr

                          Revenues: $4,090,000./yr

                          Net Costs: Net income $1,389,000./yr

The Garrett process relies heavily upon resource recovery to achieve

high performance results. It is designed to recover 90 percent of

the raw materials in the solid waste with a complex system of shred-

ders, screening, air classification and magnetic sorting processes.

The remaining material is pyrolized to effect maximum recovery of

fuel products. Estimates show a recovery rate of approximately one

barrel of good quality oil per ton of refuse and a solid material

volume reduction of 92 percent.

The Landgard Process

Developer: Monsanto Enviro-Chem Systems, Inc./EPA
                  St. Louis, Missouri

Status: 35 Ton/Day prototype successful; 1000
                  Ton/Day plant near completion in Baltimore,
                  Maryland

Recovery: Steam, char, ferrous metals, aggregates

Capacity: Input: 1000 Ton/Day @ 310 Days/Yr
                  Output: 1,500 million pounds steam/yr
                          70 tons/1000 tons input metals
                          170 tons/day glassy aggregate
                          96 percent volume reduction of
                          input

Economics: Basis: 1000 Ton/Day
                  Capital Costs: $14,742,000.
                  Operating Costs: $9.60/Ton of input
                  Revenues: $4.70/Ton
                  Net Costs: $4.90/Ton

The Landgard system is designed to dispose of all types of municipal wastes with a lesser emphasis on maximum resource recovery of raw materials. It is a complete package unit for treating influent solid wastes and the wastes generated by the process. The system entails use of an initial shredding operation which continuously feeds into a rotary kiln where exhaust gases pass over the incoming wastes to initiate pyrolysis. The gases are then treated and burned to produce steam. The hot residue char is water-quenched and the magnetic metals removed. Cooling water is clarified and recycled on site and the remaining residue is screened for glassy aggregates and transported to landfill sites. The Landgard system, discussed in previous chapters, is the only pyrolysis municipal solid waste process nearing completion on a large scale.

Animal Waste Problem

Wastes from farm animals are greatly increasing with the current rise in animal farming. The increased size of these farm operations and their geographical concentration creates large stock-piles of wastes which usually can only be disposed of seasonally. This is because manure has traditionally been used to fertilize agricultural land. However, the large quantities of manure effect high handling costs and the increased trucking distances to applica-tion sites effect high transportation costs reducing the competitive economics of manure as compared to chemical fertilizers. As with waste oils, animal wastes at one time generated a profit from their sale and not it costs to have it trucked away.

In the past animal waste management techniques have em-phasized waste disposal. But the loss of revenues from the sale of manure as a fertilizer results in a net cost to the animal feeder's operation and hinders the productivity and efficiency sought through large scale projects.

The higher agricultural efficiencies of today have generated a variety of pollution problems. Over the past few decades, development of animal operations resulted in both water and air pollution due to the accumulation of wastes. These pollutants have effected fish kills, lowered property values, and created nuisances to surrounding areas. Present animal waste handling methods allow odors to be significant problems especially while it is being spread on the soil. Animal wastes contribute to water pollution by increasing organic loads and adding excessive nutrients.

Animal waste characteristics vary from very dilute such as duck wastes which are similar to strong sewage, to semi-solid such as cattle and poultry wastes which may have moisture contents of 75 percent - 85 percent.

Wastes that accumulate at high animal population feeding operations consist primarily of fecal material. They also may include spilled feed, urinary deposits, and hair. Fecal wastes' characteristics vary with animal diet, health, feed preparation procedures, residence time of the wastes before collection on pen floors, and the wastes exposure to weather elements. These wastes, as collected, have a content of 70 percent total solids of which 95 percent are volatile solids. The organic nitrogen content of slightly over 2 percent results in a crude protein value of nearly 13 percent. The amino acids present in this material account for slightly less than 40 percent of the crude protein value, and the remaining portion consists of various non-protein nitrogen compounds. Table 82 shows an analysis of a typical animal waste.

TABLE 82 - ANALYSIS OF TYPICAL CATTLE MANURE[92]

	(Dry Matter Basis)	
	%	lb/ton
Proximate Analysis		
Crude protein	12.9	258
Fat	1.0	20
Fiber	33.1	662
Ash	5.2	104
Minerals		
Carbon	50.0	1,000
Nitrogen (Organic)	2.1	42
Phosphorus	0.4	8
Potassium	1.0	20
Calcium	1.0	20
Magnesium	0.4	8
Sodium	1.0	20
Sulfur	0.3	6
Iron	0.4	8
Amino Acids		
Alanine	0.47	9.4
Valine	0.27	5.4
Glycine	0.32	6.4
Isoleucine	0.20	4.0
Leucine	0.48	9.6
Proline	0.32	6.4
Theonine	0.26	5.2
Serine	0.25	5.0
Methionine	0.12	2.4
Hydroxyproline	0.03	0.6
Phenalanine	0.27	5.4
Aspartic Acid	0.53	10.6
Glutamic Acid	0.82	16.4
Throsine	0.18	3.6
Lysine	0.18	3.6
Histidine	0.10	2.0
Arginine	0.15	3.0
Tryptophane	--	--
Cystine/2	0.05	1.0
Diaminopimelic Acid	--	--
Total Amino Acid	5.00	100.0
Amino Acid/Crude Protein	0.39	--

Disposal of animal manures is a serious problem affected by the large volume to be handled, the nature of the waste, and the vicinity to large populations. Fly and other nuisance problems to nearby inhabitants must be controlled. Under these circumstances a satisfactory method of disposal of animal manure must be utilized. Sources and quantities of animal wastes can vary with such widely diverse activities as human beef consumption and climatological changes.

Human Meat Consumption

Animal wastes are being generated at an accelerated pace not only because of population increases in the United States, but because more beef and broilers are being consumed per capita. As seen in Figure 58 the U. S. consumption of beef and broilers has been steadily increasing while pork consumption has decreased slightly. Beef consumption has become the larger portion of the total meat consumption. At these consumption rates, for every million persons added to the U. S. population, livestock requirements will be boosted by 172,000 beef cattle, 24,000 dairy cattle, and 433,000 hogs.

Livestock Inventory

Livestock numbers vary with respect to each region in the United States. The North Central region contains 74 percent of the hogs, 42 percent of the cattle, and 39 percent of the poultry. The Sough Central and Western States contain another 41 percent of the cattle. The poultry population is more evenly dispersed across the United States. Broiler operations are concentrated in the South Central and South Atlantic states.

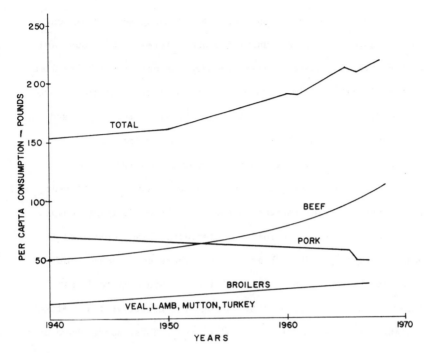

FIG. 58. U. S. consumption of meat, 1940-1967, per capita.[93]

Animal Waste Production

　　　　American farm animals generate nearly 2 billion tons of
manure annually. Estimates have been made (on a wet basis) that
about three pounds of manure is produced along with each quart of
milk and from 6 to 25 pounds of manure is produced for each pound
of weight gained by livestock.[93] A fraction of the total waste
produced remains on pastures and rangelands, but the larger portion
accumulates in feedlots and buildings and must be handled and disposed in an economical and environmentally safe manner. Each

animal specie requires different housing and management conditions
producing wastes of various quantities and natures. The amount and
characteristic of a waste is also dependent on animal size, its diet
and metabolic rate. Animals such as swine with one stomach produce
feces and urine that are similar to human wastes. Both swine and
poultry consume food which is easily digested and the amount of
excreta produced is relatively small compared to large animals.

Cattle and other ruminants produce a manure of different
quality. The bacteria inherent in the stomachs of ruminants enable
these animals to feed on cellulosic materials. Plant cellulose is
accompanied by compounds such as lignin. These compounds cannot be
digested in the rumen, the portion of the stomach used to digest
cellulose, and tend to pass right through the digestive tract effecting relatively large amounts of fecal wastes. These wastes have a
different composition from the wastes from single stomached animals.
Urinary wastes from herbivores tend to be more alkaline because their
diets are higher in compounds such as potassium, calcium, and magnesium.

Animal Waste Characteristics

As previously noted manure from different animal species will
have different characteristics. Manure from grass fed animals and
growing milk stock has lesser soil nutrients than the manure from
animals being fattened for slaughter or from work animals fed
liberally on concentrates. The biochemical oxygen demand (BOD) of
waste from steers with a diet of grain and silage can be higher
than steers feeding on grass.

Growing and milk producing animals retain more nitrogen, phosphorus, calcium, and digestible compounds from their feed for weight gain and milk production than is retained by mature stock being fattened. Table 83 gives the waste characteristics of some common livestock. Table 84 shows the analysis of some typical farm animal wastes.

Waste Treatment and Utilization

Land Application

April 9th, 1846

Planted 10 acres of manured land in corn... putting three grains to the hill, soaking in ammonia. Put a handful of compost manure, equal portions of cotton seed, swamp mud and ashes, and chaff in addition to the broadcast manure.

(From Mallorys Journal)

Animal wastes have great value as soil conditioners because of their plant nutrient and organic substance content. A more complete analysis of animal waste is given in Table 85.

There are many problems uniquely common to agricultural wastes. The management of animal manures requires a system of manure removal from its resting place, storage, transportation, treatment and disposal. Present widely accepted disposal techniques merely require application of the manure to farm land.

Collection of the manure is usually by tractor. Water is added to the manure to produce a slurry which can be readily conveyed

TABLE 83 - VARIOUS LIVESTOCK WASTE CHARACTERISTICS[93]

	Dairy Cattle	Beef Cattle	Poultry	Swine	Sheep
Animal Weight in Pounds	1400	950	5	200	100
Manure Production in Cubic Feet Per Day	1.3	1.0	0.0062	0.28	0.11
Manure Density, in Pounds Per Cubic Feet	62	60	60	62	65
Moisture,%	85	85	72	82	77
Nitrogen, % of Dry Solids	3.5	3.1	5.4	3.3	5.4
Phosphorus, % of Slurry	0.06-0.09	0.09-0.25	0.30-1.5	0.2-0.6	----
Potassium, % of Slurry	0.13-0.30	0.14-0.28	0.13-1.5	0.2-0.4	----

TABLE 84 - PARTIAL ANALYSIS OF TYPICAL LIVESTOCK WASTES*[93]

	BOD_5	Total Dry Solids	Total Nitrogen
Chickens	0.015	0.06	0.003
Swine	0.30	0.90	0.05
Dairy Cattle	1.0	10	0.40
Beef Cattle	1.0	10	0.30

*Pounds per animal per day.

TABLE 85 - TYPICAL FARM ANIMAL WASTE ANALYSIS[93]

Animal	Moisture, As A Percentage	Pounds Per Ton of Manure								
		N	P	K	S	Ca	Fe	Mg	Volat-ile Solids	Fat
Dairy Cattle	79	11.2	2.0	10.0	1.0	5.6	0.08	2.2	322	7
Fattening Cattle	80	14.0	4.0	9.0	1.7	2.4	0.08	2.0	395	7
Hog	75	10.0	2.8	7.6	2.7	11.4	0.56	1.6	399	9
Horse	60	13.8	2.0	12.0	1.4	15.7	0.27	2.8	306	6
Sheep	65	28.0	4.2	20.0	1.8	11.7	0.32	3.7	567	14

by centrifugal pumps. The manure slurry is pumped into a tank truck or into a conveying pipeline.[94]

Wastes can be applied, if the slurry viscosity is low enough, via the farm sprinkler system. Application rates from 1/3 in. to 1 in. of manure per farm acre have given good results depending on the soil requirements. Problems inherent in this application technique include large land requirements, capital equipment investment, and odor control of the spray.

Sanitary problems are lessened if the manure is immediately applied over farmland upon collection. Fresh manure can be stored only if it is made unaccessible to flies and other pests. Storage piles of solid or semisolid manure should be covered with fly-tight plastic mesh tarps or layers of soil and dry manure. The most efficient method of avoiding fly infestation is the spreading of thin layers of the manure from the feeding area on special drying beds, and tilling the area frequently. However, a drying bed of about 200 sq.ft. per cow and 1 sq.ft. per chicken is necessary and too costly for farms with large numbers of animals.

Land application of the manure recovers the nutrient content of the wastes, increasing crop production. Presently land application for disposal and nutrient recovery is not widely practiced since confinement feeding of animals has altered the practicality of land disposal of such wastes. Large quantities of waste are concentrated in relatively small areas and must be transported to disposal sites. Furthermore, difficulties occur in the application of the waste to cropland during the growing season and chemical fertilizers usually can be bought and applied more inexpensively than the utilization of free animal manure.

Aerobic Treatment

Aerobic treatment of wastes can be effected when the treatment process is not limited by the mass transfer rate of oxygen to the solution. Studies have shown that animal waste slurries, effluents, anaerobic system effluents, and run off from livestock confinement operations can be treated aerobically.

Laboratory aeration studies of hog wastes demonstrated that the BOD content of hog manure can be reduced by 50 percent to as much as 90 percent depending on the system's detention time which varied from 6 to 12 days. Longer detention times effected higher BOD removal rates. Criteria for the aeration of domestic sewage is not directly applicable to the aeration of hog manure.

The utilization of an oxidation pond for animal waste disposal is unfeasible. Oxidation ponds depend upon the oxygen given off
by algae and the oxygen transferred to the pond by natural causes.
The design criteria for these ponds ranges from 30 to 50 lb of BOD
per acre per day depending upon location. The high oxygen demand
rates of animal wastes (Table 84) would require extremely large
surface areas and volumes for adequate disposal. For example, a
confinement operation of 1,000 beef cattle would require an oxidation pond of at least 20 acres; and an operation of 1,000 hogs
would necessitate a pond of at least 5 acres.

Aerobic systems using mechanical agitation or diffused
aeration systems can be employed to reduce land requirements of
aerobic treatment. Studies utilizing an underwater air diffusion
process for aeration of a 6,500 cu.ft. pond for treating hog wastes
were made. An air diffusion rate of 35 cfm produced an effluent
with the following characteristics:

BOD	60 mg/l
COD	440 mg/l
Ammonia	40 mg/l
Nitrates	3 mg/l
DO	2.3 mg/l

Oxidation ditches can effect BOD removal rates of 90 to
95 percent, but the ditch capacity per cow and hog must be about
95 cu.ft. and 35 cu.ft. respectively, and liquid volumes of 8,000
to 10,000 gal per foot of rotor length are necessary.[94] These

land and equipment requirements of aerobic treatment methods are
too great for large scale animal waste treatment operations.

Anaerobic Treatment

        The high solids content and high oxygen demand of animal
wastes makes them readily available for treatment by anaerobic
biological systems. Anaerobic digestion of animal wastes under controlled conditions has been carried out both in the laboratory and
in the field.

        The anaerobic process has been used for many years in the
treatment of municipal sewage sludge. The primary purpose of this
process for waste treatment is the maximization of the destruction
of organics in the waste and stabilizing the sludge for disposal.
Air compressors are not necessary for anaerobic treatment since it
occurs in an oxygen defficient atmosphere thus having a very low
power requirement. Furthermore, the process produces a methane-rich
fuel gas providing plant thermal and electrical energy.

        Experiments with anaerobic processing of animal wastes have
shown that animal waste slurries with solids contents twice as high
as typical municipal sludges can be successfully treated with
fermentation times about 30 percent of municipal systems. Also,
animal waste treatment tank volumes are approximately 15 percent that
of the municipal treatment systems producing a waste with excellent
stability.

        Figure 59 illustrates a flow diagram of an anaerobic
process. Initially the wastes are mixed with water to form a slurry
which is heated to temperatures condusive to thermophilic bacteria

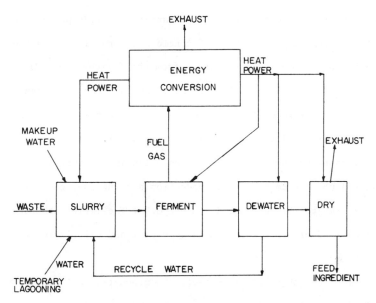

FIG. 59. Concept block diagram of use of anaerobic fermentation of animal wastes to produce a fuel gas and animal feed ingredient.

growth. It is then continuously fed into a fermentation tank equipped with agitators to constantly mix the wastes, while a number of processes take place simultaneously. The first fermentation tank process decomposes carbohydrates into simple sugars through the action of extracellular enzymes. These sugars can then be absorbed through the cell walls and enter the metabolic systems of the microorganisms. The products of the first stage of fermentation consist primarily of simple acids and alcohols, hydrogen and carbon dioxide. These products are ideal substrates for the methanogenic bacteria which generate methane and carbon dioxide. A similar degradation and synthesis process is utilized for the fats and proteins. All these

processes contribute to the reduction of original solids present in
the waste and generate a new mass of single cell protein called the
"biomass." The "biomass" is withdrawn from the fermenter and discharged to a drying bed and then used as an animal feed ingredient.

The gases generated by fermentation contain methane, carbon
dioxide and very low level of sulfur (about 0.1 percent). The combustion of these gases as fuel yields non-polluting products. The
additive reduces the cost of beef production and the fuel gas
supplies energy to the farm buildings and other facilities, thereby
saving money and eliminating the need to market and distribute the
recycle products.

The anaerobic process effects two products: a fuel and an
animal feed ingredient. Raw solid wastes are acceptable for treatment
and the process is environmentally viable for discharges with no
wastewater, solid, or gaseous pollutants.

Another anaerobic process converts animal wastes into single
cell protein animal feed. Figure 60 shows the process flow diagram.
Initially, manure is collected by mechanical or hydraulic means and
filtered to effect fiber recovery. The fiber is washed and pre-
digested through treatment with alkali and heat. The treated
fibers are then digested by microorganisms and converted to SCP.
in a one or two stage continuous operation. The resulting microbial cells are separated, washed, dried and then recycled to the
treatment tank. The major problem of this process is the difficulty
in producing a contaminant free animal feed.

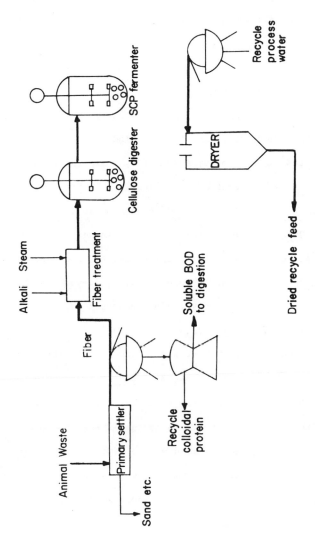

FIG. 60. Manure converting process uses predigestion before fermentation.

A pilot plant in operation in Casa Grande, Arizona, processes the wastes from 100 head of cattle. Their waste is digested
by thermophilic bacteria to produce a cellular mass of high protein
value. This mass is dried and used as a feed supplement for the
cattle. From the 350 pounds (dry basis) of manure supplied daily the
plant generates 120 pounds of feed supplement.

Lagooning as a means for waste disposal is in widespread use
for raw and primary wastewater, and for all kinds of industrial wastes.
Agricultural wastes can also be disposed of by using the lagooning
process but they must be operated anaerobically. The reason for this
is twofold: First, as previously noted large volumes of water would
be necessary to dilute manure solids content to the equivalent solids
content of sewage; and second, the land requirements for aerobic or
facultative lagoon treatment would be excessively large, also anaerobic
animal waste lagoons must be highly loaded anaerobic units.

The main difference between stabilization ponds used for
municipal sewage and industrial wastewaters and the stabilization
ponds used for the disposal of livestock wastes is that the former
treat large quantities of waters with relatively low pollutant
content, while the latter must treat large quantities of organic
solid waste matter.   The heavy loaded manure ponds generally are
operated under anaerobic conditions, making volume rather than
surface dimensions the limiting factor in their design. Other
criteria that must be considered include:

1) Odor control.

2) Fly and mosquito control.

3) Prevention of subsurface water pollution.

4) The aesthetics of the pond must be acceptable.

The following shows the volume of the stabilization pond required for each livestock animal:

Animal	Area of Stabilization Pond (ft^3/animal)
Hog	67 - 475
Chickens	6 - 15
Cattle	800 -1000

Agricultural stabilization ponds are designed with large depth to surface ratios. The influent wastes are discharged under water at the pond's center. Only a portion of the manure is converted to gas as the remaining sludge accumulates at the pond bottom. The water is maintained at designated levels with make-up water.

Experimental work at the University of California utilized 8 small manure lagoons with dimensions of 4 feet in diameter and 7 feet deep. The first year five units processed chicken manure and three units processed dairy manure. The second year swine manure was processed in the units. Manure characteristics processed in these lagoons are given in Table 86.

Feeding tests were first made on the surface liquor of the lagoon and on the manure fed. These included total and volatile solids, pH, alkalinity, electrical conductivity, and 5-day BOD.

TABLE 86 - INFLUENT LIVESTOCK WASTE CHARACTERISTICS[96]

Item	Poultry	Cattle	Swine
Total Solids (TS) in Feces and Urine (lb/day/animal)	0.066	10.44	0.795
(g/day/animal)	30.0	4,730	360
Volatile Solids (VS) (% of TS)	77.5	80.3	78.5
5-Day BOD (mgBOD/mgVS)	0.288	0.183	0.320
COD (mfCOD/mgVS)	1.11	1.0	1.2
Total Nitrogen, as N (5 of TS)	5.4	3.7	4.0
NH_4 Nitrogen (% of Total N)	74	---	75
Phosphorus, as P_2O_5 (% of TS)	4.6	1.1	3.1
Potassium, as K (T of TS)	2.1	3.0	1.4
Volatile Acids (as acetic) (% of TS)	5.8	3.3	4.8

Each week the lagoons were examined on the basis of physical appearance, scum, color, opacity of the liquor, odor, gasification, and macrobiological (fly or mosquito) activity. Bottom sludge samples were also tested.

These results showed all the lagoons to vary widely in appearance. All the results of this experiment are given in Table 87. Lagoon A was lightly loaded with poultry manure and showed characteristics similar to the typical aerobic sewage lagoon. It contained algae and maintained dissolved oxygen content in the surface waters during the daytime. Lagoon B had a pinkish color due to the photosynthetic anaerobic bacteria Thiopedea rosea. This lagoon was not visibly pleasing and the dissolved oxygen in the surface waters was negligible. All the remaining poultry lagoons had a septic and aesthetically disagreeable appearance.

The cattle waste lagoons developed a surface crust of relatively inert lignaceous hay stems and grain hulls. This crust did not harbor fly and mosquito breeding although it did support weed seed growth. These lagoons showed the most promise since the crust covered the units like the floating cover on a municipal digester. This crust is inert and can be easily broken and removed, transported, and spread on a field without problems of excessive odors or fly breeding.

The swine manure lagoons periodically developed a thin crust of liginaceous material. About 50 percent of the time no crust formed on the surface which appeared pale black and clear. In the summer months this liquor supported considerable amounts of drone fly larvae. Variations in loadings did not seem to change lagoon appearance. Well digested animal waste sludges were similar to a well digested municipal sludge. They were black, well flocculated, and dewatered, with a sweet to tarry odor. The less digested sludges from the more highly loaded lagoons had the sour odor of their respective manures and were generally lighter in color. Sludge appearance appears to be good indicator of lagoon performance.

These experiments can be summarized as follows for animals:

1) Odor is a very important criteria of the acceptability of a lagoon. Objectional odors were detected upon approaching the experimental area. This odor clung to clothing and hair much like

TABLE 87 - ANIMAL WASTE LOADING RATES AND PROCESSES[96]

Line	Item	Poultry Manure Lagoons				
		A	B	C	D	E
1	Period of Loading (days)	771	771	771	771	771
2	Loading rate (1b BOD day/cu ft)	0.000297	0.00101	0.00175	0.00300	0.000163
3	Loading rate (1b BOD day/cu ft)	0.00103	0.00391	0.00614	0.0106	0.0161
4	Loading rate (1b BOD/day/acre)	95	350	615	885	1110
5	Loading rate (cu ft/ animal)	49.8	13.6	8.35	4.82	3.44
6	General appearance of surface	Green, algae-laden, occasional few feathers	Pink to tan, feathers usually covers surface	Brown, completely covered with floating wet feathers	Brown, completely covered with floating wet feathers	Brown, completely covered with dry crust of feathers and manure
7	Characteristics of liquor	Green to yellow green, dissolved oxygen available	Rose-colored to tan, no dissolved oxygen	Brown to olive-brown, no DO	Olive-brown, no DO	Thick, olive brown, no DO
8	Gas effervescence from surface	Pin-point, in summer only	Year-round sludge lifting in summer	Year-round sludge lifting in summer	Year-round but no sludge lifting	Year-round but no sludge lifting
9	Sludge appearance	Black, well digested	Black, well digested	Brownish black, less well digested	Black, well digested	Brownish black, partially digested
10	Characteristic odor	None except occasional stagnant pond	Septic, slightly manurey, some ammonia	Sharp, putrid, manurey, some H_2S	Sharp, putrid, manurey	Very sharp, putrid, manurey
11	Odor strength (0 - 4)	0	1	4	3	4
12	Flybreeding in summer	Mosquito larvae only	Occasional larvae T. tenax	Numerous larvae T. tenax	Abundant larvae T. tenax	Abundant larvae T. tenax

tobacco smoke and remained as a residual odor on the person after he left the area.

2) Lagoons have a great appeal to many farmers because they can eliminate or significantly reduce fly problems from manure. Large numbers of flies were attracted to the fresh manures when they

Dairy Manure Lagoons			Swine Manure Lagoons		
F	G	H	J	K	L
398	398	398	232	232	232
0.00810	0.00150	0.00187	0.00153	0.00332	0.00432
0.00119	0.00830	0.0104	0.00195	0.0108	0.0137
205	350	511	388	775	1180
1830	1000	795	124	67	45
8-in.(20cm) crust, straw, hay-stems, and grain hulls, dry on top	12-in.(30cm) crust simi-lar to lagoon F	14-in.(36cm) crust, simi-lar to lagoon F	Black and opaque often with algae scum plus occasional floating grain hull	Thin mat of grain hulls frequently on surface, covered with algae scum	Wet grain hulls frequently covered surface, algae scum on this
Pale green, with manure particles dispersed, no DO	Deep green, with manure particles dispersed, no DO	Deep green, with manure particles dispersed, no DO	Clear, pale black, no DO	Clear, pale black, no DO	Clear, pale black, no DO
Masked by manure crust	Masked by manure crust	Masked by manure crust	Year-round, but no sludge lifting	Year-round, but no sludge lifting	Year-round, occasional sludge lifting in summer
Black, well digested, but full of inert hay	Black, well digested, but full of hay	Black, well digested, but full of hay	Black, well digested	Black, well digested	Black, but less digested than lagoon K
Manurey, H_2S when crust broken	Manurey, H_2S when crust broken	Manurey, H_2S when crust broken	Stagnant pond, not manurey	Stagnant pond, also manurey	Stagnant pond, also manurey
1	1	1	1	2	2
Occasional larvae T. tenax	Occasional larvae T. tenax	Occasional larvae T. tenax	Occasional larvae T. tenax	Numerous T. tenax	Numerous T. tenax

were added to the lagoons in the summer. The poultry manure fed to
the lagoons often was infested with MUSCA larvae and pupae; these
floated to the surface, and though some emergence of adults did occur,
most maggots drowned and the pupae were waterlogged.

3)   Organic Matter Destruction. The reduction of the
poultry manure total and volatile solids as shown in Table 88
indicate excellent waste biodegradation and stabilization.

Stabilization lagoons for the disposal of agricultural
animal wastes have been determined to:

1)   effect a considerable reduction in the total and
organic solids content of manure,

2)   effect a nearly complete reduction of the BOD, and

3)   cause these reductions through utilization of both
biological degradation and infiltration loss.

Animal Feed

Except for the capital investment in cattle, the cost of
feed is the largest expenditure in cattle production. A 10 percent
reduction in feed expenditures would effect a 50 percent increase in
net return to the farm operator. Therefore, a system of waste management which could utilize the nutritious values of cattle fecal matter
would be most desirable.

Animal wastes contain considerable energy and nutritive
value. The energy content of chicken feces ranges from 3.22 to 4.48
calories per gram of dry matter and the nitrogen content from 0.03 to
0.07 grams per gram of dry matter.

The utilization of dried chicken manure as a feed supplement
for chickens and ruminants can effectively reduce an operator's costs.
It can be fed to dairy cattle upon drying and the milk produced from
such cattle on this type feed was normal.

TABLE 88 - MATERIALS BALANCE DATA FOR MANURE LAGOONS[96]

Line		Poultry Manure Lagoons					Dairy Manure Lagoons			Swine Manure Lagoons		
		A	B	C	D	E	F	G	H	J	K	L
1	Total Solids Fed (lb)	138.3	406.1	662.2	943.9	1,168.6	208.1	324.3	457.0	110.0	217.7	325.4
2	Total Solids Recovered at End (lb)	34.9	98.0	304.0	190.0	362.6						
3	Degradation and Loss of TS (%)	74.8	76.0	54.2	79.9	69.0						
4	Volatile Solids Fed (lb)	99.4	229.8	506.0	723.8	898.9	156.9	249.0	353.4	84.1	168.4	251.4
5	Degradation and Loss of VS (%)	78.9	80.5	65.1	85.8	84.9						
6	Inorganic (non-volatile) Solids Fed (lb)	38.9	106.2	156.2	220.0	259.7	51.2	75.3	103.6	25.9	49.3	74.0
7	Loss (by infiltration) of Inorganic Solids (%)	64.2	62.8	18.2	60.2	45.3	26.0	42.6	61.2	26.0	52.0	79.1
8	BOD Fed (lb)	23.6	80.3	139.6	199.6	249.7						
9	Degradation and Loss of BOD (%)	96.6	96.9	92.5	96.0	94.6						
10	Volatile Acids Fed (lb)	4.2	15.5	22.7	40.0	50.6	6.7	10.6	15.1	5.2	10.6	15.9
11	Degradation and Loss of Volatile Acids (%)	87.0	94.8	97.7	95.7	96.0						
12	Nitrogen (N) Fed (lb)	5.1	18.18	33.6	48.4	61.0	7.7	12.0	16.8	4.5	9.0	13.5
13	Loss of Nitrogen (%)	76.4	74.4	61.3	79.4	60.8						
14	Phosphorus (P_2O_5) Fed (lb)	4.4	15.9	28.6	41.3	52.0	2.2	3.5	4.9	3.4	6.8	10.2
15	Loss of Phosphorus (%)	64.8	63.3	25.0	51.4	16.4						
16	Potassium (K) Fed (lb)	2.0	7.2	13.0	18.7	23.6	6.2	10.5	13.8	1.5	3.0	4.6
17	Loss of Potassium (%)	90.5	87.2	43.0	80.7	74.8						
18	Average Infiltration of Water (in/day)	2.7	1.1	1.5	2.2	2.0	2.6	1.7	1.7	1.0	0.7	0.5

Beef animals and sheep were able to utilize broiler feces as a feed supplement. The rate of gain and carcass grade were not significantly different for beef steers fed 25 percent broiler litter and the taste of the meat was unaffected. Both sheep and steers consumed a combination of cattle feedlot manure and hay and it was noted that combining such materials offers the cattle feeder an opportunity to improve feed efficiency and at the same time reduce the cost of removing manure from the feeding pends. Another study found that concentrated cattle manure could be successfully fed to pullets and laying hens. Egg production was effected only slightly.[93]

One of the important new sources of food and animal feed protein is single cell protein (SCP). SCP substrates include those derived from oil, wastes, and renewable resources as manure. Some of the advantages of SCP are, (1) the microorganisms do not depend on agricultural or climatic conditions, but are cultured in large fermentation vessels, (2) they have rapid mass growing rates, (3) genetic experimentation for protein improvement can be readily undertaken, and (4) production of SCP is not limited by land surface or sunlight.

Bacterial SCP is generally comparable to fish meal, running around 60-75 percent crude protein; yeast SCP is more like soy meal, running 45-55 percent crude protein; and mycelial fungi SCP is usually somewhat lower in protein content.

Conversion of Animal Waste to Oil

The quantities of animal wastes generated have been steadily increasing with the population growth of farm animals necessary

to feed the ever growing human population. Animal wastes make up a
large portion of the total organic solid waste produced. The quantity
of moisture and ash free solid organic waste generated in the United
States in 1971 is estimated at 800×10^6 metric tons. These solid
organic wastes have a fuel oil and fuel gas potential that is equivalent to 15 percent of the oil or 38 percent of the natural gas demands
of the United States.[97]

      There are three major processes for conversion of animal
wastes to synthetic fuels: hydrogenation, pyrolysis, and bioconversion.
Hydrogenation and pyrolysis are available processes and it is expected
that they will probably be commercialized within this decade.   Bioconversion has received very little attention and as a result, commercialization is unlikely before the mid 1980's.

      In hydrogenation, organic waste and about 5 percent alkaline
catalyst (i.e. sodium carbonate) are mixed in a vessel with carbon
monoxide and steam at initial pressures of 100 to 250 atmospheres and
temperatures of 240° to 380°F for about an hour or less.

      Under optimum operating conditions, as much as 99 percent
of the carbon content can be expected to be converted to oil or nearly
85 gallons per ton of dry waste. Normally, in practice, a conversion
rate of approximately 85 percent can be obtained. In order for this
reaction to be self-sustaining, a portion of this converted oil is
used to provide heat and carbon monoxide for the reaction. This
results in a net yield of about 50 gallons per ton of dry animal waste.

The animal waste oil is a heavy paraffinic oil with an oxygen content of about 10 percent and a nitrogen content of nearly 5 percent. Sulfur content is relatively low, about 0.4 percent, below the sulfur content limit of most cities. The heating value of this fuel oil is about 15,000 BTU per pound. In comparison, the widely used No. 6 fuel oil has a combined oxygen and nitrogen content of about 2 percent and a heating value of about 18,000 BTU per pound. The original heating value of the raw animal wastes varies from 3000 to 8000 BTU per pound.

Another method for production of synthetic fuels from animal wastes is pyrolysis. The major disadvantage of pyrolysis for small scale operations is the production of at least three different types of fuel: gas, oil, and char; thus increasing handling and marketing problems. However, pyrolysis is performed at atmospheric pressure making construction and operating costs lower than those of hydrogenation.

The pyrolysis of animal wastes involves the initial heating of the wastes to about 1000°F in a heat exchanger with no oxygen. For every ton of animal wastes, pyrolysis produces about 40 gallons of oil, 160 pounds of char, and varying amounts of a gas with a heating value of 400 to 500 BTU per scf. The gas and a portion of the char are utilized by the process to make it self-sustaining. The oil produced has an oxygen content of about 33 percent, nitrogen content of less than 1 percent, and a sulfur content of less than 0.3 percent. This high oxygen content results in an oil with a heating value of only about 10,500 BTU per pound. The heating value of this oil is actually about 75 percent that of No. 6 fuel oil because the animal waste oil

has a specific gravity of about 1.3 as compared to 0.98 for the No. 6
fuel oil.

Research into many processes to generate a synthetic fuel
from animal wastes are in various stages of development. Some have
shown promising results and it is expected that oil conversion of
animal wastes will eventually involve large scale conversion plants.
These plants will be centrally located and receive animal wastes from
large areas and in large quantities. Large amounts of oil could then
be produced to make such a plant economically feasible.

Methane Production

There are many small scale installations of manure to methane
plants. Farmers have used manure methane to heat barns, silos and
other facilities. Others have been more enterprising and run cars on
this methane.

The applications prove that methane production from animal
wastes is technically feasible. However, small scale applications do
not provide for complete utilization of the animal waste resource.
Large animal waste processing plants may in the future contribute to a
solution to the natural gas shortage problem by generating methane.

Methane may actually develop into a renewable fuel. Plant
life has always utilized energy from the sun through the process of
photosynthesis and stored it in organic carbon bearing compounds.
Death of the plant brought anaerobic bacteria to consume the decaying
plant cells and through their metabolic process generated methane, or
natural gas. Natural gas is found in underground reservoirs, or at the
earth's surface due to escaping gas from faulted rock formations.

Natural gas has for years been a by-product of sewage treatment plants. These plants usually can operate pumps, compressors,
and/or heating facilities with this gas. The anaerobic process of a
sewage treatment plant is very similar to that of a domestic septic
tank or that of an animal waste treatment plant. A slurry of sewage
and water is conveyed to an oxygen deficient tank and natural bacterial
action completes the process. The operations can become complex, since
the gas must be continuously collected and the sludge removed, and the
tank temperature maintained at 95°F for optimum process conditions.

Methane can be effected by pyrolysis and this has been previously discussed. Bioconversion of animal wastes can produce large
amounts of natural gas under the proper operating conditions.

Bioconversion of the animal wastes is a highly complex micro-
biological process. Generally, the process involves the anaerobic
digestion of the wastes to generate carbon dioxide and methane. The
organic solids of the animal wastes are initially solubilized by
enzymes into compounds that can be utilized by anaerobic bacteria.
Cellulose and starch are broken down to the simple sugars, and the
proteins are broken down to amino acids. Only the fatty acids are
not affected by these enzymes.

In the anaerobic environment the bacterial metabolic rate
is limited and is dependent on the chemical oxygen in the organic
matter being decomposed. The chemical reactions of the metabolic
process results in the formation of acids.

The high acid concentration retards bacterial growth and
the system approaches biological equilibrium. The system then develops

a new bacteria type to utilize these organic acids. These bacteria

metabolize the organic acids into carbon dioxide and methane.

Metabolism of the amino acids liberates ammonia which neutralizes

some of the remaining acids. As the pH rises bacteria growth in-

creases under the more favorable conditions producing a large methane

bacteria population. These methane bacteria permit the further de-

gradation of the more complex organic compounds.   These organic com-

pounds are finally broken down to methane and carbon dioxide by direct

metabolism. A schematic diagram of the metabolic degradation of animal

wastes is shown in Figure 61.

Operating problems of the digester are kept to a minimum

provided a balanced bacterial population of acid formers and methane

formers is maintained. The organic portion of the animal wastes upon

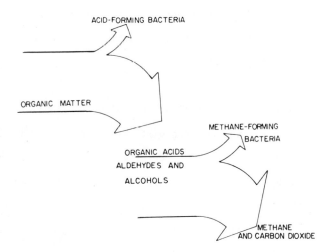

FIG. 61. Schematic representation of acid phase and methane
phase of anaerobic digestion.

addition to the digester will then be quickly converted to methane and
carbon dioxide.[94,95,96,98] A more detailed discussion of anaerobic
digestion will be given in the following chapter.

Experimental work by the Bureau of Mines installation shows
that the production of methane from manure is economically feasible.
A plant is scheduled to go into production in the near future in
which cattle manure will be biotreated to methane. The plant is to be
built in the vicinity of several cattle feedlots in the Panhandle area
of Oklahoma. The anaerobic digestion facility is to annually convert
about 90,000 tons of manure into 640 million ft^3 of methane, plus
byproduct sludge to be used as fertilizer. The gas is expected to be
sold to a natural gas distributor and become the first large scale
commercial use of bioconverted gas for interstate use. At $1.33/1,000 ft^3
the methane will be priced substantially more than the current cost
of natural gas. However, as governmental controls of the price of
natural gas diminish, the methane produced in this fashion will
become more economically attractive.

Conclusion

Alternatives to the traditional animal waste disposal method
of fertilizer land spreading have been distinctly directed toward resource recovery, particularly energy recovery. Processes to generate
an energy source from these materials have been developed and include
the following works:

1)   The Bureau of Mines utilizes anaerobic degradation of
animal wastes to produce methane.

2)   The Bureau of Mines has produced methane and char with
another process.

3) The Bureau of Mines has also produced oil from bovine manure.

4) A southwestern company is building a biogasifier plant to produce methane from cattle manure.

5) A major corporation has a process for producing single cell protein from manure for use as animal feed supplement.

The products recovered from animal wastes normally include methane gas, oil, char, and soil stabilizers. Market conditions can fluctuate to eventually make the production and sale of these substances more economically feasible. The animal wastes can then be utilized effecting another viable energy source and eliminating a pressing disposal problem.

General Information on Sanitary Landfills

The sanitary landfill as defined by the American Society
of Civil Engineers is a "method of disposing of refuse on land without creating nuisances or hazards to public health or safety by utilizing the principles of engineering to confine the refuse to the
smallest practical area, to reduce it to its smallest practical volume, and to cover it with a layer of earth at the conclusion of each
days operation, or at such more frequent intervals as necessary."

Sanitary landfilling is a method of controlled solid waste
disposal in which four basic functions are performed: (1) the landfill site is prepared to accept the municipal solid wastes; (2) the
solid wastes are deposited, spread out, and compacted into thin layers;
(3) these wastes are regularly covered; and (4) the cover material
is then compacted. The final recovered site could be developed into
recreational, storage or agricultural facilities.

The basic operation is comprised of the processes of
spreading, compacting, and covering the solid wastes. Two general

sanitary landfill methods, the area method and the trench method,
are available and currently in use. A third method, the slope or
ramp landfill, is sometimes used in combination with the area or
trench methods.

The area sanitary landfill requires the solid wastes
to be deposited on the land and then be spread and compacted by a
bulldozer or other equipment. The solid wastes are covered with a
layer of soil which is then compacted. The area method is best
applicable to flat or gently sloping terrains, and is also used to
fill land depressions such as quarries, ravines, valleys, etc.
Normally, the soil and other cover material must be trucked to the
sanitary landfill sites; sometimes it is available at the site.

The trench sanitary landfill requires a trench to be dug
in the ground and the solid wastes are deposited in it. These
wastes are spread into thin layers, compacted, and covered with the
dirt originally removed from the trench. The trench method is best
applicable to flat terrains with water tables that are relatively
deep. Under normal conditions the materials originally removed from
the trench are utilized, requiring a minimum of hauling for cover.

The ramp or slope method (a variation of the area and
trench landfills) requires the solid wastes to be deposited on the
side of an existing slope. The wastes are spread into thin layers
and compacted on the slope by bulldozing equipment. The cover
material is usually obtained just ahead of the working face and
spread and compacted on the slope. This variation of landfilling
is generally acceptable to all terrains and is commonly used with
either area or trench sanitary landfill techniques.

The cell is the building block common to each method. The solid wastes are spread and compacted in layers within a confined area. Regularly, depending on each landfill or waste type, and in most operations, daily, the solid waste is covered completely with a thin, continuous layer of soil, which is then compacted. The unit of compacted waste and soil cover is called a cell. A lift is a series of adjoining cells all of the same height and the completed landfill consists of one or more lifts. Figure 62 shows the construction of cells and lifts to make a landfill. Cell dimensions are determined by the ability of the waste to be compacted; its final resting volume depends on the density of the in-place solid waste. The density of the compacted solid waste within the cell should be at least 800 lb per cu yd. This density will vary with the waste type and will be higher if large amounts of construction wastes, glass, and well compacted inorganic materials are present. The 800 lb per cu yd will be much too high

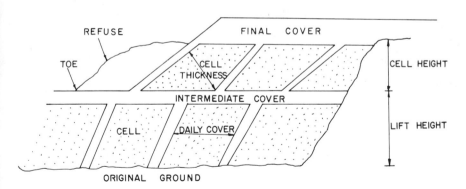

FIG. 62. Cell construction.

if the wastes are predominantly vegetation, plastics, synthetic
fibers, or rubber powder and trimmings.

Cell height varies with the availability of land and
cover material as well as solid waste composition. Cell heights
of 8 ft are common but heights up to 30 ft are common in large
operations.

Sanitary landfill requires that provisions be made for
leachate control and gas movement. Groundwater or infiltrating
surface water passing among solid wastes can produce leachate, a
solution of dissolved and finely suspended solid matter and microbial waste products. Control of leachate production and movement
to a great extent involves the prevention of water from entering
the landfill. Present leachate control techniques require careful
site selection with basic criteria as to topography and hydro-
geological conditions. Advanced concepts of leachate control include the collection and treatment and/or recirculation of contaminated waters through the fill.[99]

A number of gases are associated with the decomposition
processes within landfills. These gases include: carbon dioxide,
methane, nitrogen, oxygen, hydrogen, and hydrogen sulfide. Hydrogen
and hydrogen sulfide are not found in abundance in sanitary landfills.
Hydrogen sulfide would probably only be present in significant quantities if the infiltrating waters contained high concentrations of
sulfates, such as sea water. The predominate gases produced are
carbon dioxide and methane. Studies have estimated that over 90
percent of the gas produced by solid waste degradation in large land-

fills is carbon dioxide and methane.[100] Landfill gases must be considered in any landfill environmental impact study, because methane
can explode and inhibit the growth of vegetation, and carbon dioxide
readily dissolves in water to form carbonic acid becoming part of the
leachate problem.

Landfill gas problems can be minimized by the dispersion
of the gas into the atmosphere, or by making the cover material relatively impermeable and shunting the gases laterally. Gas movement
can be controlled by the natural soil, hydrologic, and geologic conditions of the site. Should these prove to be inadequate, methods
to control gas permeability can be constructed. There are two basic
methods of gas control by permeability: permeable and impermeable.

The permeable method utilizes gravel filled vents and
trenches which provide a path of least resistance to allow the
gases to exit. In the permeable system vent pipes are sunk through
a relatively impermeable top cover to allow the gases to escape.
Perforated collection tubes placed in shallow gravel trenches
within or on top of the solid waste feed the vertical vent pipes.
Pipe vents can also be attached to the suction end of a pump and
the landfill gases pumped out. This method has high operating and
maintenance costs.

Another impermeable cover control method regulates gas
movement by blocking the flow of gases and venting them through
the top cover. Common materials used for impermeable covers are
compacted clays and synthetic membranes.[101,102]

Composition of the Solid Wastes

The composition of the solid waste deposited in a landfill is important because ore of its components is the substrate on
which the microorganisms feed. The composition of this substrate
ultimately determines the quality and quantity of the gaseous products of the landfill. The solid wastes are comprised primarily
of various organic compounds which can be decomposed by certain
microorganisms. However, large amounts of inorganic materials such
as metals, glass, rock and other natural debris are also present and
are generally not affected by microbial action and remain as solids
in the landfill.

Solid wastes are not homogeneous in content. The amounts
and types of organic materials change from region to region and vary
with the seasons and climatic conditions. In the winter the solid
waste would contain relatively small amounts of grass and leaves.
However, grass clippings and leaves would be readily available in
solid wastes in the summer and fall seasons, and their quantities
would also vary from an urban to suburban locale.

Typical solid waste studies have become widespread and
compositions are usually specified by both weight and chemical basis.
Table 89 gives a rather complete breakdown of the typical solid waste
by weight. Table 90 gives a proximate analysis of the waste and
shows the variation in content of most components is at least 100 percent. Table 91 breaks down the organic portion of the solid waste
into its components.

TABLE 89 - TYPICAL MUNICIPAL SOLID WASTE CONTENT BY WEIGHT[103]

Component	% of Weight
Paper	42.0
Wood	2.4
Grass	4.0
Brush	1.5
Greens	1.5
Leaves	5.0
Leather	0.3
Rubber	0.6
Plastics	0.7
Oil, Paints	0.8
Linoleum	0.1
Rags	0.6
Street Sweepings	3.0
Dirt	1.0
Unclassified	0.0
Garbage	10.0
Fats	2.0
Metals	8.0
Glass and Ceramics	6.0
Ashes	10.0

TABLE 90 - PROXIMATE ANALYSIS OF A TYPICAL SOLID WASTE[104]

Measure	Minimum	Maximum	Average	
% Moisture	20	60	38	
% Carbon	8	35	24	wet basis
% Nitrogen	0.2	3.0	1.0	
BTU/Lb	3000	6000	4500	
Ash %	4	9	6.5	
Carbon %	20	50	40	dry basis
Nitrogen %	0.3	5	1.0	
BTU/Lb	6000	10000	7700	

TABLE 91 - THE ORGANIC COMPONENTS OF THE ORGANIC PORTION
OF MUNICIPAL SOLID WASTE[105]

Moisture	20.73%
Cellulose, Sugar, Starch	46.63%
Lipids	4.5%
Protein	2.06%
Other Organics	1.15%
Inerts	24.93%

Microbiology and Biochemistry of Anaerobic Decomposition

In order to understand landfill decomposition and gas
production it will be necessary to examine some basic facts of
anaerobic digestion process. Little information is available specific to the landfill decomposition process. However, the anaerobic
digestion of sewage sludge is the same process except the origin
of the organics is different.   These anaero-processes result in the
generation of the gaseous products of methane and carbon dioxide in
landfills.

Anaerobic decomposition is affected by the attack of a
mixed microbial culture metabolizing organic and inorganic material
in the absence of oxygen. The end products of organic decomposition
are carbon dioxide and methane. Since the major portion of degrad-
able landfill material is organic this is the source of nearly all
landfill gases. The end products of inorganic anaerobic decomposi-

tion are hydrogen sulfide and nitrogen. These account for only a small portion of the landfill gases generated.

The decomposition of the organic material evolving methane and carbon dioxide is a two stage process. The first stage is the acid production stage. Extracellular enzymes initiate this phase by breaking down the complex, long chained organic compounds to smaller, soluble components to be utilized by the bacteria. The organic substrate (i.e. food scraps, leaves and other vegetation, paper, etc.) is composed of carbohydrates, proteins, cellulose, and fats. The enzyme amylase breaks down the carbohydrates to glucose. The enzyme protease breaks down the proteins to amino acids. The enzymes cellulase and cellobiase break down cellulose to glucose. The enzyme lipase breaks down the fats to smaller chained fatty acids. During this process the bacteria expend more energy than is recoverable. These processes prepare the substrate for bacteria metabolic reactions and generate no gas.

These products of enzyme activity are now available for utilization by the bacteria. The bacteria metabolize the glucose, amino acids and fatty acids to organic acids (primarily acetic and propionic), and by products of carbon dioxide, hydrogen, ammonia, and H_2O.[106]   The amounts and types of organic acids produced vary and are not readily ascertained.

Acetic acid is the most prevalent organic acid intermediate found in the methane fermentation of fats, carbohydrates, and proteins and about 70 percent of the methane produced in the subsequent anaerobic fermentation results from its degradation.[107]

The following is a list of microorganisms responsible for the generation of organic acids.[108]

> Bacillus cereus
>
> Bacillus knelfelkampi
>
> Bacillus megaterium
>
> Bacteriodes succinogenes
>
> Clostridium carnofoetidum
>
> Clostridium cellobioparus
>
> Clostridium dissolvens
>
> Clostridium thermocellulaseum
>
> Pseudomonas formicans
>
> Ruminocossus flavefaciens

It has yet to be determined as to whether the microbial reactions
during this stage are the results of facultative or obligate anaerobes.
Facultative microbes can grow either in the presence or in the
absence of molecular O_2. Obligate anaerobes can grow only in a totally
oxygen deficient atmosphere.[109] These microbes are more tolerant
and less susceptible to environmental changes and disturbances than
those of the next methanogenic stage.

The second stage of organic decomposition generates methane
and carbon dioxide. This is called the methanogenic stage and
equations summarizing several of the reactions include:

$$CH_3COOH \dashrightarrow CH_4 + CO_2 \qquad\qquad \text{(acetic acid)}$$

$$4CH_3CH_2COOH + 2H_2O \dashrightarrow 7CH_4 + 5CO_2 \qquad \text{(propionic acid)}$$

$$2CH_3CH_2CH_2COOH + 2H_2O \dashrightarrow 5CH_4 + 3CO_2 \qquad \text{(butyric acid)}$$

$$2CH_3CH_2OH \dashrightarrow 3CH_4 + CO_2 \qquad \text{(ethanol)}$$

$$CH_3COCH_3 + H_2O \dashrightarrow 2CH_4 + CO_2 \qquad \text{(acetone)}$$

The equations show that fermentations of acetic, propionic, and butyric acids, ethanol, or acetone all result in the same products. The nature of the end products is independent of the original substrate structure. Van Niel published his carbon reduction theory in 1938 which is generally accepted today.[110] The basis of this theory is that the organic compounds fermented by methane bacteria are completely oxidized to carbon dioxide, and this oxidation process is followed by a reduction process in which some or all of the carbon dioxide is reduced to methane. The Van Niel fermentation of acetate is as follows:

Oxidation	$CH_3COOH + 2H_2O \dashrightarrow 2CO_2 + 8H$
Reduction	$8H + CO_2 \dashrightarrow CH_4 + 2H_2O$
Net	$CH_3COOH \dashrightarrow CH_4 + CO_2$

The microbes that effect this stage are a highly specialized physiological group of bacteria commonly referred to as the "methane producing" bacteria. The methane bacteria are highly susceptable to disturbances in their environment and are very sensitive to oxygen, more so than most other anaerobes. Even a trace of oxygen can destroy the bacteria within a few minutes.[109] Another methane bacteria characteristic is that they possess a highly specialized metabolic system which produces methane as a major product. This ability to metabolize methane is not a common trait to all anaerobic bacteria but is restricted to a specialized group.

As a group, methane bacteria not only specialize with respect to the chemical aspects of the metabolic end products, but are highly specific in the types of substrates they can utilize. Furthermore, each species characteristically is restricted to the use of a few compounds. Table 92 gives some species of methane bacteria and their corresponding utilizable substrates.

The methane bacteria show a rather extreme development of substrate specificity. This effects the need for several species of methane bacteria to complete the fermentation of the many organic compounds present in municipal solid wastes. Three species of methane bacteria are necessary just to complete the fermentation of a simple compound such as valeric acid. Valerate is initially oxidized by Mb syboxydans to acetate and propionate which cannot further be processed by this organism. Methane is then formed in a coupled reduction reaction. A second species, Mb. propionicum, metabolizes the propionate to acetate, carbon dioxide, and methane, and cannot further reduce the acetate. Therefore, a third species such as Methanosarcina

TABLE 92 - METHANE BACTERIA SPECIES AND THEIR UTILIZABLE
SUBSTRATES[99]

Species of Methane Bacteria	Oxidizable Substrates
Methanobacterium formicium	H_2, CO, formate
Methanobacterium omelianski	H_2, ethanol, primary and secondary alcohols
Methanobacterium suboxydans	Butyrate, valerate, caproate
Methanosarcina barkerii	H_2, CO, methanol, acetate

methanica is required to ferment the acetate. A balanced population
of bacteria must be established to develop a culture capable of
causing a rapid and complete fermentation of complex mixtures of organic compounds.

Another factor important to the growth of methane bacteria
is the hydrogen ion concentration. These bacteria grow best in the
pH range from 6.4 to 7.2. Below pH 6 and above pH 8 the growth rate
of methane bacteria dramatically drops.[110]

Table 93 gives ten species of methane bacteria which metabolize organic compounds and the corresponding substrate reactions.

TABLE 93 - ORGANIC METABOLIZING METHANE BACTERIA SPECIES
AND THEIR CORRESPONDING SUBSTRATE REACTIONS[108]

Organisms	Reactions
Methanobacterium Soehngenii Methanococcus Mazei Methanosarcina Methanica Methanosarcina Barkeri	$CH_3COOH \longrightarrow CH_4 + CO_2$
Methanobacterium Propionicum Methanococcus Mazei Methanosarcina Methanica Methanobacterium Suboxydans	$4CH_3CH_2COOH + 2H_2O \longrightarrow$ $7CH_4 + 5CO_2$ $2CH_3(CH_2)_2COOH + 2\ H_2O \longrightarrow$ $5CH_4 + 3CO_2$ $2CH_3(CH_2)_2COOH + 2H_2O + CO_2 \longrightarrow$ $CH_4 + 4CH_3COOH$
Methanobacterium Omelianskii	$2CH_3CH_2OH \longrightarrow 3CH_4 + CO_2$ $2CH_3CH_2OH + CO_2 \longrightarrow CH_4 +$ $2CH_3COOH$
Methanobacterium Suboxydans	$CH_3COCH_3 + H_2O \longrightarrow 2CH_4 + CO_2$

Nitrogen and hydrogen sulfide gases can also be generated by anaerobic decomposition. The microbial process of denitrification produces nitrogen from the reduction of nitrate ions while they are electron acceptors. The denitrification process begins immediately upon the depletion of molecular hydrogen.

Hydrogen sulfide is produced by the bacteria genus De-sulfovibrio, which oxidizes substrates such as sodium lactate or hydrogen gas. Relatively small amounts of hydrogen sulfide are generated in sanitary landfills, but because of its characteristic pungent odor, these quantities are undesirable. Also, hydrogen sulfide is easily solubilized in water resulting in a highly corrosive solution. On the other hand, hydrogen sulfide can minimize the toxic effects of soluble copper, zinc, nickel and the heavy metal salts to anaerobic decomposition.[106,111,112]

Environmental Factors Affecting Methane Gas Production

The anaerobic digestion process in which methane gas is generated is the most complex and sensitive of all the biological waste processes. The methanogenic stage of anaerobic digestion is highly susceptible to environmental changes. Methane production is affected by many environmental factors including:

Moisture Content

The methanogenic bacteria are obligate anaerobes. Oxygen in any amount destroys their activity, however, they form spores and as anaerobic conditions return their activity is stimulated again.

pH

The optimum pH range for anaerobic digestion is approximately 6.6 to 7.6. The non-methanogenic or acid forming bacteria
can tolerate a wide range of pH from 5 to 8.5 as opposed to the near
neutral pH conditions required by the methanogenic bacteria.

Alkalinity

In order to maintain the pH near neutral conditons a buffer
capacity of about 1,000 mg/l as $CaCO_3$ must be available to maintain
the pH near neutral conditions and inhibit the accumulation of the
volatile acids produced by the non-methanogenic bacteria preventing
an acidic environment. An ammonium ion concentration in excess of
100 mg/l expressed as ammonia is considered to be in excess.[106,114]

Oxidation Reduction Potential (ORP)

The production of methane requires the oxidation reduction
potential of the bacteria environment to be well into the negative
range and generally less than 200 mV. This is necessary for decomposition in both soil and anaerobic digestors.[106]

Temperature

Anaerobic digestion can take place in almost any reasonable ambient temperature, but the rate of digestion varies greatly
with the temperature and the rate of temperature change. At a
temperature of 55°F (13°C), about 90 percent of the desired digestion is effected in about 55 days. Increasing temperatures, increase
the bacteria metabolism rates such that at 75°F (24°C) the digestion
time is cut to 35 days, at 85°F (29°C) to 26 days, and at 95°F (35°C)
to 24 days.

Digestion in the temperature range between 80°F (27°C) and 100°F (38°C) is called mesophilic digestion, favorable to mesophilic organisms. Thermophilic bacteria require temperatures above 100°F to cause thermophilic digestion. Satisfactory results are obtained from both organism types but thermophilic are selective to the digestion of specific organic solids. The temperature for digestion must be consistently maintained to within 2 to 3°F since the microorganisms cannot tolerate sudden or large temperature changes.[115]

Nutrition

The nitrogen to carbon ratio in the substrate of anaerobic bacteria should be approximately 0.06; a ratio below this level would result in little anaerobic activity. Nitrogen is not easily obtained by methane bacteria in the presence of other anaerobic and facultative organisms.

The ionic concentration which would effect the most efficient anaerobic digestion includes: 0.01M for the monovalent ions, sodium, potassium, and ammonium; and 0.005M for the divalent ions, calcium and magnesium.

Soluble copper, zinc, and heavy metal salts are toxic to anaerobic organisms. However, under anaerobic condtions most toxic heavy metals can be neutralized by an equivalent concentration of sulfides.[112,116,117]

The previous discussion noted some of the basic environmental factors affecting the anaerobic digestion of the sludge to generate methane in order to form a basis for methane production in a landfill environment. The major difference in the two processes is that the anaerobic digestion of sludge is very sensitive and is therefore

operated under contained and controlled conditions and that landfill

anaerobic decomposition is not controlled or suitable for control.

Pattern of Gas Production in Sanitary Landfills

A typical gas production process in a landfill can be

described providing two basic assumptions are made: First, after

the solid waste is placed in the landfill no aeration occurs. Second,

environmental conditions within the landfill are sufficient to en-

courage and sustain anaerobic digestion.

Methane gas production in a sanitary landfill involves

four stages: Stage I - Aerobic; Stage II - Anaerobic Non-Methan-

ogenic; Stage III - Anaerobic Methanogenic Unsteady; and Stage IV -

Anaerobic Methanogenic Steady. Figure 63 is a graphical representa-

tion of these four stages.

Stage I occurs at the time of placement of the solid waste

in the landfill, as the oxygen present in them is utilized effecting

aerobic decomposition. The reaction for complete decomposition of

the solid wastes under aerobic conditions with the empirical formula

$CH_aO_bN_c$ is as follows:

$$CH_aO_bN_c + 1/4(4+a-2b-3c)\ O_2 \longrightarrow CO_2 + 1/2(a-3c)\ H_2O\ +\ cNH_3$$

Calculations at 20°F and at 1 atmosphere pressure show

(a) 14.5 cubic feet of O_2 to be utilized and 14.5 cubic feet of

CO_2 produced in the aerobic decomposition of one pound of carbo-

hydrate; (b) 17.6 cubic feet O_2 utilized and 17.1 cubic feet CO_2

produced in the decomposition of one pound of protein; and

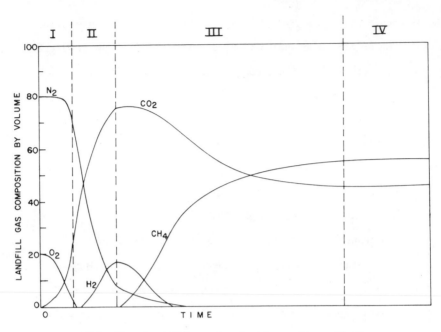

FIG. 63.   Landfill gas production pattern phase.

(c)   35.2 cubic feet O_2 utilized and 24.8 cubic feet CO_2 produced
in the decomposition of one pound of fat.

The Stage II process of anaerobic activity begins after
the oxygen supply is depleted. During this period the maximum concentration of carbon dioxide occurs and hydrogen production begins.
Nitrogen is simultaneously displaced and then produced by a denitrification process. The lag in methane production after the anaerobic
process begins is probably due to the need for adequate amounts of
CO_2 to act as a hydrogen acceptor. This is similar to the acid
formation stage of methane fermentation previously discussed.

Studies at landfill sites have shown the gases generated
to have peak CO_2 concentrations of 70 percent within 11 days of
deposition, indicating a rapid decomposition of certain carbohydrates and other readily decomposible materials. In similar studies
maximum CO_2 concentrations of 95 percent after 45 days in a simulated
test cells were observed.

Stage II hydrogen concentrations are approximately 20 percent by volume.

Stage III is characterized by increasing concentrations of
methane until its generation stabilizes. The methanobacteria become
increasingly active and use up the available landfill hydrogen. Carbon
dioxide and nitrogen concentrations are also reduced.

The completion time for all three stages varies widely
with each solid waste. Time intervals of 180, 250, and 500 days
have been observed.

During Stage IV the gas production rate and composition
should remain constant. Landfill gas compositions of this stage
have a methane content of 50 to 70 percent and a carbon dioxide
content of 30 to 50 percent.

The reaction for the complete anaerobic decomposition of
solid wastes is given in the following empirical equation:

$$CH_aO_bN_c + 1/4(4-a-2b+3c)H_2O \longrightarrow 1/8(4-a+2b+3c)CO_2 + 1/8(4+a-2b-3c)CH_4$$

From the above equation calculations can be made to show that (a) the
metabolization of one pound of carbohydrates produces 7.33 cubic feet

CO_2 and 7.3 cubic feet CH_4;   (b) the metabolization of one pound
of protein produces 8.3 cubic feet CO_2 and 8.8 cubic feet CH_4;   (c) the
metabolization of one pound of fat produces 7.2 cubic feet CO_2 and 17.6
cubic feet CH_4. The major decomposition process in a landfill is
anaerobic.

These stages of decomposition do not occur in succession,
but usually in combinations, where one or two processes become domin-
ant depending on environmental conditions and then another one takes
over, etc. For example, should oxygen be made available to the solid
waste by external means, Stages III and IV anaerobic decomposition
will cease and Stage I will begin.[99,106,108,118,119,120]

A Mathematical Treatment of Landfill Decomposition Rates

A mathematical model of landfill decomposition rates, and
therefore gas generation, has been developed based on landfill operating characteristics. The model predicts landfill decomposition by
using a two step calculation which assumes a constant decreasing
rate of gas generation preceded by a constant increasing rate of gas
generation. The development of gas generation equations from these
assumptions resulted in a landfill life span of three phases, each
represented by a different equation.

Phase One - filling of landfill represented by growth and
exponential decay:

$$\theta(t) = 1/k \ (e^{-kt} + kt - 1)$$

Phase Two - increasing decay rates represented exponentially:

$$\theta(t) = e^{-kt} - 1$$

Phase Three - decreasing decay rates represented exponentially:

$$\theta(t) = e^{-kt} (k-1) + 1$$

where $\theta(t)$ = % decomposition

k = constant

t = time

These equations were evaluated and based on landfill gas generation data, estimates of landfill volume and elapsed time for gassing. For phase one 0.19 decomposition was assumed for a period of twelve years. This yielded a k value of 0.0038. In a similar manner, values of k of 0.0199 for the second phase and 0.029 for the third phase were obtained. Substitution of the values of the constant into the appropriate equations resulted in the following landfill characteristics:

1) Half life occurs at 23 years

2) 90.5 percent decomposition occurs at 80 years

3) 99.1 percent decomposition occurs at 160 years

The constants were obtained from data taken from a very small sample and can not justifiably be called "typical." However, with development, this method could become a viable and valuable tool for gas generation predictions and availability.[99]

Factors Influencing Gas Production in a Landfill

Moisture

The gas production rate of a solid waste increases with increased moisture content with maximum gas evolution rates occurring at moisture contents from 60 to 80 percent wet weight. The methane content of the gases produced also increased with increased moisture

content. Studies have shown the methane concentration to vary from
almost negligible in the landfill existing without water to greater
than 50 percent in the saturated landfill. Results of the moisture
content of landfill material of various ages and depths taken from
5 New York area landfill sites **are** shown in Table 94. Solid waste
moisture content increases with age and also with depth since water
flows naturally downward. However, the percentages are well below
those most conducive to increased activity. These results indicate
that possibly the moisture content of many landfills is so low
that anaerobic functions are minimized evolving lesser amounts
of gas. Suggestions have been made to irrigate landfills to
increase the moisture content and gas rates, but the possibility of leachate problem is produced.[99,106,121,122]

TABLE 94 - MOISTURE IN SAMPLES EXTRACTED FROM 5 NEW YORK AREA
LANDFILLS[123]

Age (Months)	Top (2' to 4') %	Middle (5' to 7') %	Bottom (8' to 10') %
0-1	18.9	20.9	22.8
3-6	19.2	23.8	20.9
6-9	21.7	24.3	28.4
9-12	24.5	26.7	33.5
12-18	25.2	25.9	31.7
18-24	25.7	30.3	34.3
24-30	20.9	24.1	28.3
30-36	25.5	28.1	32.2
36-48	24.0	28.1	32.4
48-120	21.1	29.5	33.4
360-420	20.9	22.9	21.3

On the other hand, excess infiltration of water may
hamper methanogenic activity. Others have observed decreases in
the methane content of gases evolved from a field test cell after
being infiltrated by large amounts of water generated from melting
snow and ice. A temperature drop of less than 3°C occurred, but
this was not interpreted to be the sole factor of decreased gas
generation, since these lower gas rates were also noted with large
amounts of infiltration without a change in refuse temperature.

Field cell tests in Sonoma County near San Francisco
demonstrated similar results. These tests were performed specifically to determine the effect on solid waste stabilization of
applying, under various operating conditions, excess water, septic
tank pumpings, and recycled leachate to a sanitary landfill. Data
obtained for the first two years of this 3 year project are presented
in Table 95 and Figure 64. The control cell A shows minimal gas
production; control cell C, with a continuous application of water,
shows minimal methane production, but the carbon dioxide concentration has been maximized. It is theorized that the added water
had a substantial O_2 concentration which could be toxic to the
anaerobic organisms present in the landfill.

Temperature Effects

Temperature is of importance in the degradation process
because it is directly related to the rate of biological activity
in the landfill. Temperatures of landfill operations usually
follow a similar pattern. After deposition of the solid waste

TABLE 95 - OPERATIONAL MODES FOR SONOMA COUNTY FIELD TEST CELLS[123]

Cell Designation	Liquid Conditioning	Initial Liquid Used	Operation Daily Liquid Application (gal/day)	Liquid Used	Purpose of Cell
A	none	none	none	none	control cell
B	field capacity*	water	none	none	to determine the effect of high initial water content on refuse stabilization
C	none	none	700± (200-1000)**	water	to determine the effect of continuous water throughflow on leachate character
D	none	none	1000± (500-1000)	recirculated leachate	to determine the effect of continuous leachate recirculation on leachate character
E	field capacity*	septic tank pumpings	none	none	to determine the effect of high initial moisture content, and septic tank pumpings on refuse stabilization

* Field capacity is the condition when a sufficient quantity of fluid has been added to the refuse to cause a significant volume of leachate to be produced.

** Range of variation in daily application of fluid.

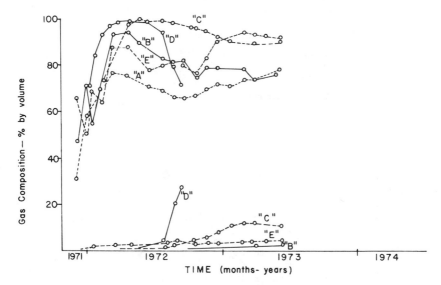

FIG. 64.   Sonoma County field cell tests.

a peak temperature is reached as a result of aerobic decomposition.
This rise in temperature is also dependent on the original placement temperature, and this is shown in Figure 65, a plot of solid
waste placement temperature versus Stage I peak temperatures. The
maintenance of aerobic conditions yields sustained high temperatures;
however, the conversion to anaerobic decomposition will reduce the
landfill temperature significantly. Test solid waste cells operated
in parallel, under aerobic and anaerobic conditions, resulted in a
maximum temperature of 193°F in the aerobic cell after 48 days and
after 76 days the temperature was still a high 174°F. The peak
temperature in the anaerobic cell of about 120°F was reached within
60 days of placement and temperatures then declined to a low of 90°F.

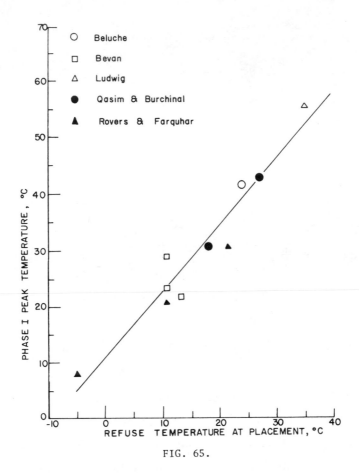

FIG. 65.

There are many problems associated with landfill investigations because of the interrelationship and interdependence of the many parameters involved. However, some of the results obtained can be summarized as follows:

Placement of solid wastes under winter conditions increased the initial moisture content of the solid waste and greatly inhibited bacterial activity.

.   During the winter the surface of the solid wastes
became and remained frozen until the spring. Anaerobic activity
was unable to increase temperatures enough to permit active decomposition.

.   The top 10 to 15 feet of the solid waste was seen
to undergo biannual reversals. These were assumed to have a major
effect on the rates of microbial activity.

.   The surface layers of frost and snow created during
the winter inhibited atmospheric exchange with the solid waste
causing the system to turn anaerobic. This can result in H_2S
production, CO_2 and CH_4 increases and a gas pressure buildup. The
gas composition could have detrimental affects on root growth and
the pressure increase could force the gas to move laterally beyond
the landfill site.

Optimal landfill temperatures for anaerobic gas production
are about 90-95°F. Lower temperatures effect lower gas production
rates. Therefore, the gas generated in many landfills will probably
be less than maximum, especially those in colder climates, and seasonal
variations will produce fluctuations in gas production rates.

pH

From our previous discussion on methane production the
optimal pH for gas generation is near 7.0. Other pH conditions prove
unfavorable to the methanogenic bacteria, reducing methane gas production. The non-methanogenic bacteria can tolerate more acidic
conditions, about pH 5.2. Supporting data for the applicability of
these conditions for landfills is scarce.

An analytical approach to predict acid concentrations in a landfill by parametrically plotting the log of the two major factors of temperature and moisture content against the reciprocal of the acid concentration was developed. The resulting parametric surfaces were approximated mathematically and computer models were developed. These models could predict the following information concerning landfills:

1) Acid concentrations per unit weight of fill material.

2) Concentration-time curves for any acids formed under any conditions.

3) Theoretical parametric acid surfaces.

4) Superimposed plots of predicted and observed values of acids to facilitate the evaluation of the theoretical equations.

As more data becomes available it will be possible to develop a comprehensive program to predict acid and gas production in a landfill.[99,108,124]

Other Factors

Many other parameters affect methane generation including toxic materials, solid waste composition, compaction and cover material, oxidation-reduction potential, topography and hydrogeology.

Soluble copper, zinc, and nickel salts are toxic to anaerobic decomposition. These materials may be present in some industrial wastes which are dumped at municipal landfills destroying anaerobic bacteria.

With a varying organic material content, i.e., carbonaceous, proteinaceous, fatty, and cellulitic, the gas content
and production rate also vary. Carbonaceous materials yield
three times as much gas as proteinaceous materials, but in three
times as long. A carbon to nitrogen ratio of 16:1 is considered
to be optimal for methane production during sewage sludge digestion.
Therefore, gas production from the decomposition of paper with a
carbon to nitrogen ratio of approximately 400:1 will differ from that
of grass at a carbon to nitrogen ratio of 16:1.

Refuse placement and cover procedures can effect gas
migration. The microbial population present in the adjacent soil
can effect gas production. The topography and hydrogeology of the
site are important since they are critical factors in governing
water infiltration.

The generation of methane gas in a landfill site is
dependent on a large number of factors as we have seen in this
discussion. Each of these factors is capable of significantly
affecting gas composition and rates. Due to the nature of a landfill most of these factors are beyond human manipulation.[99]

Energy Recovery Potential

Estimating the potential energy (the amount of methane)
available in a landfill is a difficult if not impossible task. As
noted previously, the structural, physical, and chemical character-

istics of landfills are infinitely variable. To estimate the
methane potential the following parameters must be known:

      1) The type and extent of biological decomposition
(i.e. the amount of methane generated).

      2) The amount of methane which has escaped or is
escaping.

      3) The rate of methane generation.

Since each of these considerations is affected by a multitude of
factors it is highly probable that no two landfills would exhibit
exactly the same gassing patterns.

Recovery Process

      The gaseous products of decomposition consist almost
exclusively of carbon dioxide and methane. The landfill gases
content of methane and carbon dioxide vary with the progressing
stages of decomposition; however, for a reasonably progressed stage
of decomposition the respective percentages are about 50:50 as seen
in Figure 66. This gas mixture has a heating value of approximately
500 BTU/ft^3 and could be used on site to run either a gas turbine or
reciprocating piston engine generator. The electricity could be
used locally or supplied to landfill facilities. The nonrecoverable
costs for this operation would be minimal providing the engine/gen-
erator did not require permanent installation. The disadvantage of
this arrangement would be a low thermal efficiency estimated to be
25 percent or less. The gas could also be treated for removal of the
carbon dioxide to raise the heating value to approximately 1000 BTU/ft^3
making the gas of pipeline quality.

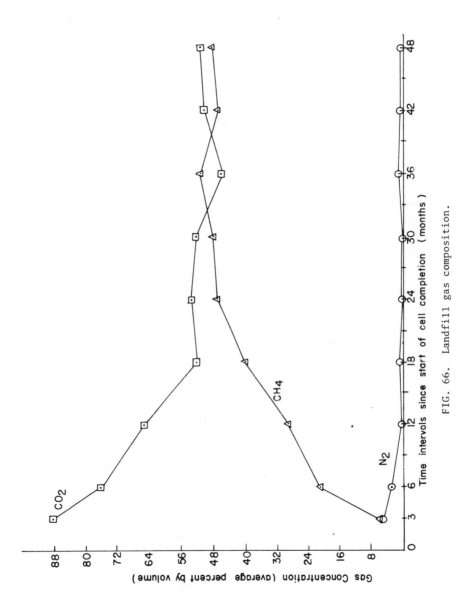

FIG. 66. Landfill gas composition.

NRG NuFUEL of California has developed a process for
using a molecular sieve to remove high concentrations of CO_2 from
gas streams. The molecular sieves are hydrated metal aluminum
silicates possessing unique structures with centrally located
cavities whose only entrances are apertures of a specific size.
The unique structure of the molecular sieves allows them to be
used to separate gases or liquids selectively and with respect to:
the diameter and configuration of the molecules, the degree of the
unsaturation of the molecules, and the polarity of the molecules.
Molecular sieves can effect these separation processes by means of
their apertures or "pores," which can screen out large or irregularly
shaped molecules, and by means of their preferential attraction for
polar molecules.

The NRG gas purification process operates by circulating
the contaminated gas through three vessels packed with adsorbent
molecular sieve. The gas is pumped into the first vessel where
impurities are adsorbed by the molecular sieve until the capacity
of the bed is exhausted. The stream is then switched to a second
vessel, while the first vessel is regenerated by depressurization
and evacuation. When the second vessel is exhausted, the stream
is switched back to the first vessel, while the second is regenerated. This "pressure swing cycle" is repeated continuously on a
controlled basis.

Over a period of time, the molecular sieve slowly loses
its capacity to adsorb all impurities, requiring each vessel to be
removed periodically from the pressure swing cycle and heated to

drive off residual contamination. This is called the "thermal swing cycle." A third vessel is employed to permit the process to continue without interruption. These cycles revolve continuously, with each vessel taking a turn through the thermal swing cycle, while the other two alternate on pressure swing cycles. The net result of this procedure is a continuous flow of pipeline quality methane gas.[99]

This type of methane purification process would require permanent installation with a high capital and operating cost partly attributable to the increased power requirements for a pressurized drop through the system. Provisions must be made for accomodating the generated methane gas whether or not it is to be ultimately utilized for energy purposes. Presently, it is assumed that the gas has to be either vented or flared. This apparatus can be extensive as a pumping well.

To extract methane from a landfill the logical method is from a well. In order to obtain the negative gas pressure gradient necessary for extraction a pump is also required. For a continuous operation the volume of gas extracted will ultimately have to be equal to the volume of the gaseous products of decomposition per unit time. The zone of influence of the negative pressure gradient caused by the pumping will determine the volume of gaseous products extracted. Limited test well data are available on pumping rates and the corresponding zones of influence. Table 95 shows the data from a forty foot deep landfill, for the optimum extraction rates.

It should be noted that upon initial operation of a well there is a considerable supply of gaseous products stored from time past.

After a certain period of time this backlog of gaseous products will be exhausted and extraction of a higher volume of gases than generated by decomposition will only serve to draw air into the landfill and cease anaerobic activity. High extraction rates probably can be sustained in continual operation. Therefore, the calculation of the maximum possible volume of gaseous products capable of sustained extraction for a well is made. First calculating the stoichiometric products of decomposition of 1 lb of solid waste using the average composition previously cited and the corresponding metabolized products for anaerobic decomposition the following is obtained:

	%	$CO_2 (ft^3)$	$CH_4 (ft^3)$
Carbohydrates	.4748	3.480	3.466
Fats	.045	.324	.792
Protein	.0206	.171	.181
		3.975	4.439

Therefore, for each pound of solid waste in the landfill, 3.975 ft^3 of CO_2 and 4.439 ft^3 of CH_4 are generated over the entire decomposition cycle.

Using the mathematical model of gas decomposition rates for a 23 year half life, and assuming constant decomposition over this period (not necessarily true) we obtain per pound of solid waste:

$$CO_2 \quad \frac{1.987 \ ft^3}{23 \ year} = .0002367 \ ft^3/day = .000000164 \ ft^3/min$$

$$CH_4 \quad \frac{2.220 \ ft^3}{23 \ year} = .0002644 \ ft^3/day = .000000184 \ ft^3/min$$

Calculating volume of influence assuming a radius of influence of a 94 ft pipe inlet at bottom of well, and a 40 ft deep landfill:

Volume of influence = $3.14 (94)^2$ x $40 = 1110364.5 ft^3$

Calculating total pounds of refuse in this volume based on 800 lb/yd^3 compaction:

$$800 \ lb/yd^3 = 29.63 \ lb/ft^3$$

Total weight of refuse in zone of influence:
$29.63 \ lb/ft^3$ x $1110364.5 \ ft^3 = 32899689. \ lbs$

Multiplying by volumetric availability:

CO_2 32899689. x .000000164 $ft^3/min/lb = 5.39 \ ft^3/min$

CH_4 32899689. x .000000184 $ft^3/min/lb = 6.04 \ ft^3/min$

Therefore, it can be seen that there is a notable disparity between the calculated volumetric availability of gas and the quoted optimum extraction rate. Initial well operations would serve only to collect gases already generated over long periods of time. However, sustained operations are not possible since actual gas production rates controlled by biological action are incompatible with withdrawal rates. Even if the zone of influence (94 ft) was entirely occupied with gas, at the optimum extraction rate of 190 cfm the gas supply within the zone would be depleted in approximately 4 days. For steady state gas withdrawal the volume of influence would have to be increased by a factor of 16.6. This would correspond to a radius of influence of 383 ft which would be equivalent to a surface area of approximately 11 acres. It must be remembered that this is a conservative estimate since no loses were assumed in the generation of the gas. NRG requirements are for a landfill to be viable for gas extraction it must be capable of producing 1,000 ft^3/min. Using this criteria a landfill would have to be a minimum of 55 acres and 40 ft deep with no losses.

From this it is concluded that the potential for sustained extraction of methane from a landfill for significantly large use is not possible in other than extremely large landfills with ideal conditions for biological decomposition and containment of gases. However, the possibility of extracting and utilizing the methane on site for local power generation purposes is possible and could be developed with little problem.[99]

Wood Wastes Availability and Potential

Advocates of alternate energy sources have begun to
examine the wood processing industry as an untapped energy reservoir. Each year tremendous quantities of timber are harvested across the country to be processed into lumber, plywood, or
pulp. These wood operations generate large volumes of wood and
bark residues. Bark, chips, edgings, sawdust, sanding dust,
trimmings are byproducts of primary manufacturing. These by-
products are comprised of more than half of the volume of the
original log. Some utilization of these residues has been
accomplished through reprocessing into pulp, pulp products,
composition board, and sometimes fuel.

Raw material for pulp plants in Oregon is supplied
almost entirely by the wood residues of sawmills and plywood
plants. Eastern U.S. manufacturers of pulp generally use round
wood and effect a bark residue as it is stripped from the log
before processing. Wastes from the logging operations are almost entirely neglected except for disposal concerns. Tables 96,97

TABLE 96 - ESTIMATE OF FOREST RESIDUES PRODUCED IN THE
UNITED STATES ANNUALLY (BILLION CUBIC FEET)

Source of Residue	Volume	
	Available	Possibly Recoverable
Residue from logging operations	3.5	3.0
Fire, disease and insect damaged timber	4.5	2.0
Unused, precommercial thinning	1.5	1.0
Total	9.5	6.0

TABLE 97 - AVAILABLE WOOD RESIDUES IN THE PACIFIC COAST
STATES ANNUALLY[126]

In Thousands of Dry Tons

State	Logging Residue	Mill Residue
Alaska	375	
California	5,355	3,168
Oregon	4,764	2,990
Washington	3,800	1,113
Total	14,294	7,271

give the amount of unused residue annually available in just the U. S.
and the Pacific Coast States.

Logging wastes are those wood wastes effected by timber
harvesting operations. These wastes are generated in appreciable
amounts. In the national forests of western Washington and Oregon

the gross volume of logging residues averages about 57 dry tons
per acre. In the Ponderosa pine region or eastern sectors of these
states, the average gross volume is only about 5.0 dry tons per
acre. Thus, Table 98 shows the available quantities of logging
residues per acre of lands logged to/from region to region even
within states. The useable volume of these materials normally
ranges from 69 to 89 percent.

Milling wastes are those wood wastes generated during
the processing of the logs into products. Usually these quanti-

TABLE 98 - LOGGING RESIDUE QUANTITIES OF THE PACIFIC NORTHWEST
STATES[126]

State, Region, and Ownership	Logging Residue		
	Gross Weight (Dry Tons Per Acre)	Net Weight (Dry Tons Per Acre)	Percent
Oregon and Washington Douglas-Fir Region			
National Forest	56.9	39.1	69
Other Public	33.1	25.3	76
Private	18.6	16.6	89
Ponderosa Pine Region			
National Forest	4.5	3.6	80
Private	5.0	4.4	88
California (All Regions)			
National Forest	18.3	14.6	80
Private	23.8	17.1	72

ties can amount to 1-1/4 dry tons per 1000 feet of lumber cut and
1/2 dry ton per 1000 feet by 3/8 inch plywood produced. In mills
in the Pacific northwest nearly 23 million dry tons of these materials are produced annually with 7 million dry tons going unused.
Table 99 characterizes the unused portion of milling wastes.

Other wood residues that are potentially available
include: 1) Fire damaged trees; 2) Insect and disease ridden
trees, 3) Driftwood, 4) Construction wood wastes, and 5) the
wood portion of municipal solid wastes.

TABLE 99 - MILLING WOOD WASTES UNUSED IN THE PACIFIC NORTHWEST
STATES[126]

Location and Source Industry	Wood			Bark
	Coarse[1]	Medium[2]	Fine[3]	
California				
Veneer and Plywood	73.1		6.2	113.0
Lumber	822.9	309.4	786.6	1,053.2
Other			2.7	1.0
Oregon				
Veneer and Plywood	243.7		35.3	434.3
Lumber	462.8	176.3	601.8	908.5
Shake and Shingle	27.8		59.8	40.0
Washington				
Veneer	89.0		4.3	103.5
Lumber	134.9	91.1	170.7	231.7
Shake and Shingle	69.0		124.6	94.1
Total	1,923.2	576.8	1,792.0	2,979.3

1. Veneer trim, cores, and panel trim from plywood plants; slabs, edgings,
   and trim from sawmills.

2. Planer shavings.

3. Sanderdust from plywood plants; sawdust.

These wood and bark residues can have widespread usage
in a variety of applications. The demand for products which can
utilize wood residues as a raw material is continuously rising.
Depending on the residue size and quality, they can be used in
the manufacture of plywood, lumber, paper and pulp, and building
boards.

Figure 67 shows that the demand for roundwood in the
U.S. will increase from about 12,000 cu.ft. in 1970 to nearly
20,000 cu.ft. in 1980. Table 100 breaks this demand down into
individual requirements: Increases of from about 25 percent in
saw log consumption to over 100 percent in pulpwood consumption

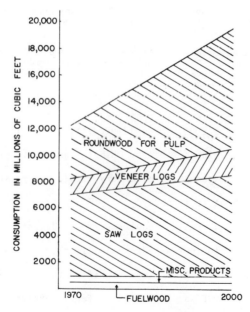

FIG. 67. The increasing domestic demand for roundwood through
2000.[126]

TABLE 100 - UNITED STATES CONSUMPTION OF ROUNDWOOD AND MILLING WOOD
WASTES THROUGH 2000[126]
(In Millions of Cubic Feet)

Product Group	Reported Consumption,1970		Projected Consumption,2000	
	Roundwood	Mill Residue[1]	Roundwood	Mill Residue[1]
Saw Logs	6,100	--	7,600	--
Veneer Logs	1,200	--	2,000	--
Pulpwood	4,000	1,600	9,100	2,800
Miscellaneous Products[2]	400	600	400	1,200
Fuelwood	500	700	500	700
Total	[3]12,200	2,900	[3]19,600	4,700

Based on medium level of projected demand and prices 10 to 30 percent above
1970 levels.

1. Residue developed in converting roundwood to such primary products as
    lumber and plywood. It is used chiefly for pulp, particle board, and
    fuel.

2. Includes roundwood for cooperage, poles, piling, mine timbers, shingles,
    and some board products; it also includes the residue used primarily for
    building boards.

3. The totals include an estimated 7,900 (1970) and 10,400 (2000) million
    cubic feet of sawtimber. The balance in each case is roundwood from
    such sources as cull and dead trees, trees less than 5.0 inches in
    diameter, and logging residue.

are expected by the year 2000. Miscellaneous roundwood products

such as poles, piling, mining timbers, shingles and board products

and wood for fuel are expected to have very little change in their

demand in 2000 over what it is today.

With increased housing requirements the domestic demand

for lumber is expected to increase dramatically through 1980 after

which the demand increase will be less significant. Domestic pro-

duction of lumber in 1970 was 34.7 billion feet. Exports of lumber

are not expected to change. The accelerated construction pace and

continued rise in the manufacture of wood products will result in
plywood and building board usage to more than double and triple,
respectively, by 2000. 36.8 billion square feet of plywood, and
about 9 billion square feet of particle board, hardboard, and in-
sulation board were consumed in 1970. Therefore, the more virgin
wood used, the more wood wastes generated. A significant portion
of the increased composition board demand is expected to be satisfied through the use of wood residues as raw materials.

Wood wastes are also expected to help satisfy a 2.4 fold
increase in paper, cardboard and pulp products. 72 million dry tons
of softwood chips were used by the paper and pulp industry in the
United States in 1972. At the present time this industry uses
wood wastes for about 30 percent of its raw materials. Wood
wastes supplied about 76 percent of the Pacific coast pulp industry's
raw material requirements.

Thus, wood residues are available and will continue to
be available throughout this century. They have been utilized to
some extent, and are expected to have increased usage in the future. However, their utilization is somewhat dependent on the consumer industry's proximity to the waste reservoir. The paper and
pulp industry of the Northwest can readily use the waste wood resource because large quantities of milling and logging residues
are generated in this region. Other locations across the United
States which house paper and pulp and composition board companies
do not have these wood residues as a cheap and readily available
raw material. For these plants to utilize wood wastes in product

manufacturing, the additional handling and transportation costs
must be considered before any long term prediction about wood
waste usage can be finalized.

Although these wood and bark residues can and do have
widespread usage in a variety of applications, the ultimate de-
termining factor in their usage is cost. The economics may prove
to be unfeasible for a relatively small milling plant to reuse or
have its wood wastes reused. Furthermore, the large plants which
generate huge volumes of wood residues may not have an outlet to
accept these wastes and may find they have to transport long
distances to be utilized. When the utilization of wood and bark
residue becomes uneconomical, a disposal problem is created. The
use of these materials as a fuel may prove to be a viable alternative.

The 7 million dry tons of wood and bark wastes generated
on the Pacific coast could replace conventional fuels at milling
facilities for their processing and operational heat requirements.
Table 101 gives some characteristics and the heating values of
typical wood residues. Figure 68 shows the moisture content of
some western woods and barks in pounds per cubic foot.

Before examining wood residue characteristics any further,
their units of measure are discussed to familiarize the reader with
the terminology.

1) Cubic Foot - The volume of logs or chunks are typically expressed in solid cubic feet. The gross cubic feet is the total

TABLE 101 - DENSITIES AND HEATING VALUES OF SOME TYPICAL PACIFIC
COASTAL WOOD RESIDUES[126]

Kind of Wood or Bark	Density (Green Volume Basis)		Moisture Content (wet basis)	Heating Value
	Ovendry Weight	Green Weight		
	--Pounds Per Cubic Ft.--		Percent	BTU per dry Lb.
Douglas-Fir:				
Wood	28	38	26	8,900
Bark	27	38	29	9,800
Western Hemlock:				
Wood	26	50	48	8,400
Bark	29	51	43	9,400
Ponderosa Pine:				
Wood	23	44	48	9,100
Bark	21	24	12	9,100
True Firs:				
White Fir Wood	23	47	51	8,200
Western Redcedar:				
Wood	20	--	--	9,700

volume of solid material and the net cubic feet term describes the
portion of the total that is considered usable.

2) Tons (Wet) - The moisture content of the wood residues
is included in this term. As seen in Figure 68, the moisture content
of wood can be as much as 50 percent by weight. When converting from
tons (wet) to tons (dry) a water content of 50 pounds per cubic foot
of wood and bark is often assumed. This is often referred to as the
wood's green weight.

3) Tons (Dry) - To measure the weight of wood residues
of varying densities and water contents the bone dry weight of the

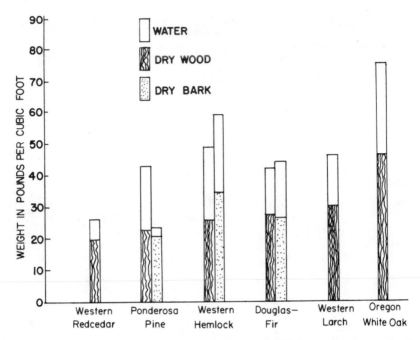

FIG. 68. Moisture content and dry wood weights of some typical woods.[126]

material is used. It is the basic unit of wood residue weight measurement.

    4) Unit - Wood wastes in the form of particles such as sawdust, hogged fuel, or chips are usually measured in bulk. A unit of wood residue is the amount contained in a volume of 200 cubic feet. Units vary with respect to solid wood content which is dependent on wood type, moisture content, and compaction. This concept is clearly shown in Figure 69 where a 2000 pound unit is less than half solid matter.

    5) Cunit - A cunit is 100 cubic feet of solid wood.

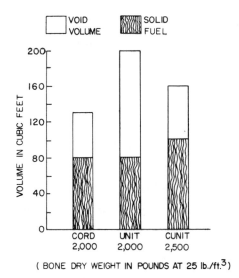

FIG. 69.   Some common weights and volumes of wood residue
measurement.

What has become evident is that wood residues vary in size

and composition. Sizes of wood residues range from dust particles

generated by sanding to large chips and slabs. The moisture content

of the sander dust is relatively small as opposed to the moisture

content of some woods which exceeds 50 percent. These wood wastes

also vary with respect to appearance.

To utilize wood residues for fuel or any application, they

must first be reduced to a size that can be handled conveniently.

Large volumes of wood and bark residues can be effectively reduced

in size by a "hog" machine. The particles produced by this piece

of equipment are commonly referred to as hogged fuel or hog fuel.

Hogged fuel can include any portion of wood or bark residues in a

reduced particle size. Additives to the hogged fuel may also include sawdust and wood shavings and ground bark. Hogged fuel may
be a conglomerate of many types of wood wastes.[126]

Wood Waste As A Fuel

          Hogged fuel, particularly in the northwestern region of
the United States, is usually sold in bulk volumes, or units. A
unit of hogged fuel roughly contains a ton (dry) of wood material.
This dry weight can vary from 2600 pounds for hogged Douglas fir
bark to 1900 pounds for sawdust to 1200 pounds for Douglas fir
shavings.

          FIG. 70. The Morbark Total Chiparvestor picks up entire trees
and reduces them to wood chips in seconds. The wood chips will go
to heating plants and other wood fiber users. (Courtesy Morbark
Industries, Inc.)

FIG. 71. To aide Michigan Technological University in their
forestry experiments, Morbark Industries, Inc. donated a Morbark
Model 22 Total Chiparvestor. Michigan Tech's Total Chiparvestor
will convert tree tops and commercial thinnings into wood chips
that will be diverted to wood fiber users and to heating plants.
Examining the Morbark Total Chiparvestor is Michigan Tech's
President, Dr. Raymond Smith (left), State Rep. (Mich.) Russell
Hellman (center), and Dr. Eric Burdo, Dean of Tech's School of
Forestry. (Morbark Industries, Inc.)

The moisture content of a particular hogged fuel varies

with the wood type, season, size of a hogged wood particle, and

whether the original logs were dry or wet handled. Bark from

Douglas fir logs that were handled in ponds contains moisture

in quantities almost equal to its dry weight. A unit of hogged

Douglas fir bark from logs handled in such a manner might have a
wet weight of approximately 5200 pounds.

The quantities of the various wood residues generated
by the manufacture of lumber and plywood can be estimated by the
conversion factors given in Tables 102 and 103. These are average
values for the estimation of wood wastes frim Oregon mills only,
since these factors vary from mill to mill depending on parameters such
as log size, quality, and species; manufacturing equipment; final
product; and the mill's quality control. Residue factors for each
mill at a particular location must be determined based on that mill's
operating conditions.

TABLE 102 - AVERAGE CONVERSION FACTORS FOR ESTIMATING RESIDUES FROM
THE MANUFACTURE OF A THOUSAND BOARD FEET OF LUMBER IN
OREGON

Item	Solid Volume[1]	Proportion By Volume	Dry Weight	
			Western Oregon	Eastern Oregon
	Cu Ft	Percent	Tons	Tons
Coarse Wood Residue[2]	43	26.0	0.580	0.516
Sawdust	22	13.4	0.297	0.264
Planer Shavings	16	9.7	0.216	0.192
Total Wood Residue	81	49.1	1.093	0.972
Bark Residue	19	11.5	0.285	0.228
Lumber	65	39.4	0.878	0.780
Total Log	165	100.0	2.256	1.980

1.   Equivalent undried solid volume.
2.   Includes slabs, edgings, and lumber trim.

TABLE 103 - AVERAGE CONVERSION FACTORS FOR ESTIMATING RESIDUES
          DEVELOPED FROM THE MANUFACTURE OF A THOUSAND SQUARE
          FEET OF EQUIVALENT 3/8 INCH PLYWOOD (ROUGH BASIS)
          IN OREGON

Item	Solid Volume[1]	Proportion By Volume	Dry Weight
	Cu Ft	Percent	Tons
Log Trim	3.4	4.4	0.046
Cores	3.7	4.8	0.050
Undried Veneer[2]	18.5	24.1	0.250
Dried Dust[3]	6.5	8.5	0.088
Sander Dust	1.6	2.1	0.021
Total Wood Residue	33.7	43.9	0.455
Bark Residue	8.8	11.5	0.132
Plywood	34.3	44.6	0.463
Total Log	76.8	100.0	1.050

1. Volumes are based on equivalent undried solid volume.
2. Undried veneer residue includes veneer clippings, roundup,
    and spur trim.
3. Dried veneer residue includes dry veneer loss and panel trim.

        Estimates of wood residues from Oregon sawmills and
plywood operations that were used for fuel in 1967 amounted to
4 million dry tons or about 27 percent of the total residues
generated. The heat content of those residues used for fuel was
70 x 10^{12} BTU which is equivalent to the heat from the total sales
of the Northwest Natural Gas Company in 1967. Western Oregon and
parts of Washington are supplied natural gas by this company.

        In the Tennessee Valley Authority region there are 844
wood waste producing plants including furniture and veneer factories, pulp and lumber mills etc. Table 104 shows the total tonnage
of wood residue produced by each industry in the region. Tables 105

TABLE 104 - BREAKDOWN BY INDUSTRY OF WOOD RESIDUES PRODUCED IN
              THE TENNESSEE VALLEY[127]

	Total Tonnage Produced (in 1,000)	Percent Unused
Sawmills	1,863	69.1
Planing Mills	109	4.3
Hardwood Dimension and Flooring	131	2.3
Special Product Sawmills	50	1.0
Veneer and Plywood	44	1.5
Prefabricated Building and Mobile Homes	12	0.8
Wooden Containers	43	2.0
Miscellaneous Wood Products (Mill Work, Cooperage, Turnings, Etc.)	181	7.4
Furniture	313	8.8
Pulp and Paperboard	669 (bark)	2.7
Boats, Sporting Goods, and Games	3	0.1

TABLE 105 - RESIDUES FROM LOGGING AND LAND CLEARING, TENNESSEE VALLEY[127]

Residues From:	Softwood	Hard Hardwoods	Soft Hardwoods	Total
	------------Thousand Tons--------------------			
Sawlogging				
Based on 1970 Data	557	2,571	509	3,637
Pulpwooding				
Based on 1973 Data	1,401	1,101	221	2,723
Strip Mine Clearing				
Based on average of				
1970-73	112	440	88	640
Other Land Clearing				
Average Annual Rate	1,199	2,810	947	4,956
Total Produced	3,269	6,822	1,765	11,956
Total Unused	3,046	6,215	1,628	10,889

and 106 show the wastewood generated from logging and land clearing

in the Tennessee Valley and the amounts of wood residues that are

unused. The total amount of wood residues produced from land clear-

ing and industrial operations in 1970 was over 12.3 million tons.

The heating value of this residue is equivalent to 4.9 million tons

of coal. Thus, availability of these wood residues is of such mag-

nitude, and considering the cost of disposal, that their utiliza-

tion as a fuel or fuel supplement is becoming imperative. Wood at

one time was the primary fuel of the country, but because of its

relatively low heating value as compared to coal and oil, it dropped

from use. However, rising gas,oil, coal and disposal prices has

resulted in renewed interest in wood as a fuel.

Fuel Properties of Wood and Bark

The basic elemental composition of most wood species is

relatively constant and this is true of bark also. Table 107 gives

TABLE 106 - SUMMARY OF WOOD RESIDUES IN THE TENNESSEE VALLEY[127]

Unused Residues From:	Softwood	Hard Hardwood	Soft Hardwood	Total
	------------Thousand Tons-----------------			
Industrial Plants	407	872	132	1,413
Logging Operations	1,947	3,620	720	6,287
Land Clearing Operations	1,100	2,594	908	4,602
Total Unused Annually	3,454	7,086	1,762	12,302
Coal Equivalent	1,272	3,032	613	4,917

If coal is worth $20 per ton, the 4.9 million tons of residue equals
$98,000,000.

If #2 fuel oil is worth 40 cents per gallon, the 4.9 million tons of
residue is worth $310,000,000.

TABLE 107 - TYPICAL ULTIMATE ANALYSIS OF BARK FUEL ON A DRY-WEIGHT BASIS

Component	Douglas Fir Bark	Western Hemlock Bark
	Percent	Percent
Hydrogen	6.2	5.8
Carbon	53.0	51.2
Oxygen	39.3	39.2
Nitrogen	0.0	0.1
Ash	1.5	3.7

the typical ultimate analysis of Douglas fir and western hemlock

barks. Hogged bark from a sawmill, which contained some quanti-

ties of wood (normally found in hogged fuel), was used for the

analysis. Fourteen other eastern Canadian species had barks with

a similar composition as shown in Table 107.

Bark and wood fuels have negligible sulfur contents and

unlike most heavy oils and coals, they do not effect sulfur com-

pound air pollution problems.

Air pollution problems due to ash are also relatively

low. Ash is the portion of the fuel which is noncombustible and

normally must be separated from the combustion gas before they are

exhausted out a stack. Wood contains small amounts of ash in the

order of less than 1 percent by dry weight. The noncombustible

content of the wood fuel is often increased as a result of dirt

and sand adhering to the bark during harvesting and handling. Ash

contents for the bark of some western wood species is given in

Table 108. Ash contents of softwood barks can range from 0.6 per-

cent for sugar pine to 2.5 percent for Engelmann spruce. Hardwood

barks had ash contents ranging from 1.5 percent for paper birch
to 10.7 percent for white oak, and coniferous eastern Canadian
barks range from 2.0 percent for jack pine to 4.2 percent for
tamarack.

From Tables 107, 108, and 109, it can be seen that bark
generally has a higher fixed carbon and ash content than wood and
usually has less volatile matter.

The heating value is one of the most important properties
of a fuel. The heating value of most wood species is about the
same, 8300 BTU per pound, provided they are moisture and resin free.
Woods with resin contents have higher heating values than nonresinous
woods because the heating value of the resin is nearly 17,000 BTU per
pound. Table 110 gives the heating values for wood and bark of some

TABLE 108 - BARK ASH CONTENTS OF SOME WESTERN WOOD SPECIES (DRY WEIGHT)

Species	Ash Content, Dry Weight Bark Percent
Douglas Fir	1.2 - 2.2
Douglas Fir	---
Douglas Fir	1.5
Douglas Fir	---
Western Hemlock	1.7
Western Hemlock	3.7
Western Hemlock	---
White Fir	2.6
Ponderosa Pine	0.7
Lodgepole Pine	2.0
Sugar Pine	0.6
Red Alder	3.1
Red Alder	2.4
Engelmann Spruce	2.5
Western Larch	1.6

TABLE 109 - TYPICAL ANALYSES OF WOOD, DRY[128]

	Percent By Weight						Heating Value BTU Per Lb.		Atmos.Air At Zero Air lb/10° BTU	CO$_2$ At Zero Excess Air, %
	Carbon C	Hydrogen H$_2$	Sulfur S	Oxygen O$_2$	Nitrogen N$_2$	Ash	Higher	Lower		
SOFTWOODS**										
Cedar, White	48.80	6.37	-	44.46	--	0.37	8400*	7780	709	20.2
Cypress	54.98	6.54	-	38.08	-	0.40	9870*	9324	712	19.5
Fir, Douglas	52.3	6.3	-	40.5	0.1	0.8	9050	8438	719	19.9
Hemlock,Western	50.4	5.8	0.1	41.4	0.1	2.2	8620	8056	705	20.4
Pine, Pitch	59.00	7.19	-	32.68	-	1.13	11320*	10620	702	18.7
Pine,Pitch White	52.55	6.08	-	41.25	-	0.12	8900*	8308	722	20.2
Yellow	52.60	7.02	-	40.07	-	1.31	9610*	8927	709	19.2
Redwood	53.5	5.9	-	40.3	0.1	0.2	8840	8266	707	20.2
HARDWOODS**										
Ash, White	49.73	6.93	-	43.04	-	0.30	8920*	8246	709	19.5
Beech	51.64	6.26	-	41.45	-	0.65	8760*	8151	728	20.1
Birch, White	49.77	6.49	-	43.45	-	0.29	8650*	8019	714	20.0
Elm	50.35	6.57	-	42.34	-	0.74	8810*	8171	717	19.8
Hickory	49.67	6.49	-	43.11	-	0.73	8670*	8039	712	19.9
Maple	50.64	6.02	-	41.74	0.25	1.35	8580	7995	719	20.3
Oak, Black	48.78	6.09	-	44.98	-	0.15	8180*	7587	713	20.5
Red	49.49	6.62	-	43.74	-	0.15	8690*	8037	711	19.9
White	50.44	6.59	-	42.73	-	0 24	8810*	8169	713	19.8
Poplar	51.64	6.26	-	41.45	-	0.65	8920*	8311	715	20.0

*Calculated from reported higher heating value of kiln-dried wood assumed to contain eight percent moisture.

**The terms hard and softwood, contrary to popular conception, have no reference to the actual hardness of the wood. According to the Wood Handbook, prepared by the Forest Products Laboratory of the U.S. Department of Agriculture, hardwoods belong to the botanical group of trees that are broad leaved, whereas softwoods belong to the group that have needle or scale like leaves, such as evergreens; cypress, larch and tamarack are exceptions.

TABLE 110 - HEATING VALUES FOR WOOD AND BARK OF SOME WESTERN SPECIES

| Species | Higher Heating Value, Per Dry Pound | |
	Wood	Bark
	BTU	BTU
Douglas Fir	-	9,400[1]
Douglas Fir	9,200	10,100
Douglas Fir	8,860	-
Douglas Fir	8,800	10,100
Douglas Fir	8,910[2]	-
Western Hemlock	-	8,900[1]
Western Hemlock	8,500	9,800
Western Hemlock	8,000	-
Western Hemlock	8,620	-
True Firs	8,300	-
(White Firs)	8,000	-
Ponderosa Pine	9,100	-
Ponderosa Pine	9,140	-
Lodgepole Pine	8,600	-
Lodgepole Pine	-	10,760
Sitka Spruce	8,100	-
Engelmann Spruce	-	8,820
Western Larch	-	8,750
Western Redcedar	9,700	8,700
Western Redcedar	9,700	
Redwood	9,210	-
Red Alder	8,000	-
Red Alder	8,000	-
Red Alder	-	8,410
Oregon Ash	8,200	-
Bigleaf Maple	8,410	-
Bigleaf Maple	8,400	-
Black Cottonwood	8,800	9,000
Oregon White Oak	8,110	-

1. Hogged bark as obtained from a mechanical debarker had some wood included.

2. Sawdust.

western species. The barks generally, because of high resin contents, have higher heating values than woods.

Wood fuel is frequently sold by the volume unit rather than weight. Table 111 gives the heating values of some species

TABLE 111 - A COMPARISON OF THE DENSITIES AND HEATING VALUES ON A
VOLUME BASIS OF SOME WEST COAST SPECIES OF WOOD

Kind of Wood	Density Dry	Higher Heating Value, Dry	Higher Heating Value	Relative[2] Heating Value
	Lb Per cu ft[1]	BTU Per Lb.	BTU Per cu ft[1]	
Douglas Fir	28	8,900	249,000	1.00
Western Hemlock	24	8,500	204,000	0.82
Ponderosa Pine	24	9,100	218,000	0.88
Lodgepole Pine	24	8,600	206,000	0.83
Sitka Spruce	23	8,100	186,000	0.75
Western Redcedar	19	9,700	184,000	0.74
Redwood	24	9,210	221,000	0.89
Red Alder	23	8,000	184,000	0.74
Black Cottonwood	20	8,800	176,000	0.71
Bigleaf Maple	27	8,400	227,000	0.91
Oregon Ash	31	8,200	254,000	1.02
Oregon White Oak	37	8,110	300,000	1.20

1. Volume based on green condition, solid volume.
2. Relative basis, Douglas fir equal 1.0, different fuel moisture
   not considered.

based on volumetric units of dry wood. These values may vary depending on the moisture content of the wood.

The amount of water included with the dry wood is also an
important property of wood and bark fuel. In order to evaporate
moisture, heat is required, which is then lost to the evaporating
moisture. Therefore, moisture has a negative heating value. Table
112 illustrates the relationship of bark moisture to heat content.
As the moisture content is increased the heat content is proportionately decreased.

The moisture content of these wood materials can be expressed in several manners. The forestry industry usually expresses

TABLE 112 - RELATIONSHIP OF BARK MOISTURE TO HEAT CONTENT[128]

Moisture - Percent	BTU Per Lb.	Lb. Water Per Lb. Dry Substance
0	8750	0.00
20	7000	0.25
40	5250	0.67
50	4375	1.00
60	3500	1.50
70	2625	2.30
80	1750	4.00
90	875	9.00

moisture in terms of a ratio of water weight in a material to its
dry weight. This ratio is usually expressed as a percentage. A
material with equal weights of dry substance and water would have
a moisture content of 1:1 or 100 percent (dry basis).

The fuel and combustion industry expresses moisture as
the ratio of water weight to the total weight. A material with
equal weights of water and dry substance would have a moisture
content of 1:2 or 50 percent (wet basis). The following equations
can be used to convert from wet to dry basis or vice versa:

$$M.C. \ (wet) \ = \ 100 \ M.C. \ (dry)/(100 + M.C. \ (dry) \ )$$
$$M.C. \ (dry) \ = \ 100 \ M.C. \ (wet)/(100 - M.C. \ (wet) \ )$$

where M.C. = moisture content in percent.

The moisture content of wood and bark fuels varies widely.
Hogged fuel generally has a moisture content of around 50 percent
(wet basis) most of which is contained in the cellular structure of
the wood.[126,127,128,129,130]

Combustion of Wood and Bark

      In the combustion of wood and bark three processes occur, consecutively at first, but then simultaneously:

      1) First, heat must be supplied to evaporate the water in the wood fuel in order to effect combustion.

      2) Volatile hydrocarbon gases are then evolved and mixed with oxygen giving off heat.

      3) More heat is released and combustion completed with the reaction of oxygen with the fixed carbon at high temperatures. Initially these processes occur in succession, but as heat is generated the wood eventually begins to sustain its own combustion and all processes occur at once.

      The stochiometric air requirement of a combustion process is that amount of air necessary to burn the carbon and hydrogen in the fuel completely to carbon dioxide and water. The Douglas fir bark fuel analysis of Table 107 is used to calculate stochiometric air requirements for its combustion. About 6.5 pounds of air is necessary to completely burn 1 pound of dry fuel. Excess air quantities for the complete burning of 1 pound of dry Douglas fir bark fuel with up to 100 percent excess air are illustrated in Figure 72.

      Depending on the moisture content, stochiometric air requirements vary and naturally so does the weight of the stack gases evolved. Figure 73 gives the weight of the stack gases evolved from the complete combustion of 1 pound of dry Douglas fir bark with various moisture contents and quantities of excess air. As the temperature of the stack gases increases their volume increases. Based on the

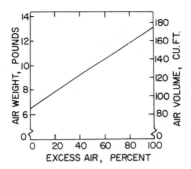

FIG. 72. Excess air requirements for complete burning of 1 lb
of dry Douglas fir bark fuel. Volume in cubic feet is calculated
for standard air at 70°F with a density of 0.075 of a pound per
cubic foot.

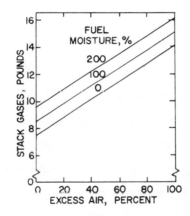

FIG. 73. Weight of stack gases evolved from the complete
combustion of 1 lb of dry Douglas fir bark fuel.

burning of 1 pound of dry Douglas fir bark with 40 percent excess

air Figure **74** shows the volume of stack gases as a function of stack

temperature and wood moisture content. 40 percent excess air is a

representative air requirement of a boiler plant burning wood fuel.

Figures **72, 73** and **74** could be utilized by a wood power plant designer

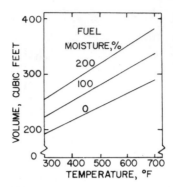

FIG. 74. The volume of stack gases affected by the combustion
of 1 lb of dry Douglas fir bark fuel with 40% excess air.

to determine equipment criteria such as fan size, ductwork, and stack
design.

Wood Burning Plants

Many firms commonly use wood and bark residue fuels. These
residues are also burned for home heating in stoves, furnaces and
fireplaces. The heat of combustion of sander dust is used in veneer
and wood particle drying processes and for the production of steam.
The latter process effects the largest industrial use of wood and
bark fuels. Steam is produced for heating, processing and power
generation through electricity. Hogged fueled steam plants range
in steam capacity from 10,000 pounds per hour for a small plant to
over 500,000 pounds per hour. A steam power plant to utilize wood
and wood products for fuel would typically consist of the following:

1) A mill or hog to process large wood scraps into a size
more readily acceptable for combustion;

2) Wood storage facilities to meet peak steam demands and
during interruption of wood fuel deliveries;

3) Equipment to control the flow of fuel into the furnace;

4) A furnace for the combustion process;

5) A boiler for steam generation;

6) Air pollution control devices;

7) A network of conveyors or other systems to conveniently handle the wood fuel and resultant ash;

8) Controls to effect automatic operation of the system;[131]

9) A fuel predrying process may be necessary for the combustion of wood residues with high moisture contents.

Probably the most common hogged wood fuel burning process involves the use of a dutch oven. Figure 75 shows a two-stage dutch oven furnace. In the first stage (dutch oven), water in the wood is evaporated and the fuel is gasified. In the secondary furnace combustion is completed. The system is gravity fed as the hogged fuel enters the dutch oven from above and forms a conical pile. The

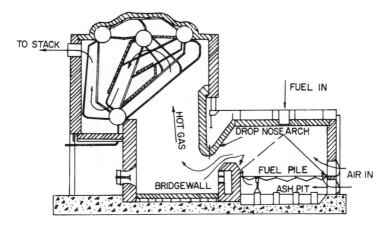

FIG. 75. Two-stage dutch oven furnace.

dutch oven was in widespread use until about 25 years ago. Presently, more efficient and larger capacity systems are available.

The fuel cell process is also a two-stage furnace as
illustrated in Figure 76 . The hogged wood fuel enters the primary
furnace compartment from overhead. The gravity fed system allows
the fuel to drop onto a water cooled grate. The wood is first gasified in the primary stage and the gases pass into a secondary com-

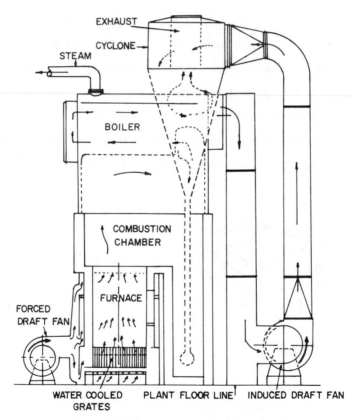

FIG. 76. Fuel cell process steam plant for wood and bark residue
fuels.

bustion compartment for complete combustion. Hogged fuel boilers
of this type are in rather widespread use throughout the western
United States. Labor operating costs are low because these steam
plants are highly automated. Low operating pressures of about 25 psi
gives these plants steam capacities ranging from 10,000 to 30,000
pounds per hour. When the moisture content of the fuel is above 100
percent (dry basis), hogged fuel dryers are necessary. Many of these
wood steam plants have applications in drying kiln processes.

Many steam plants recently constructed to be wood or bark
fueled are the spreader-stoker type. In the spreader-stoker system,
a pneumatic or mechanical spreader feeds the hogged fuel from above
onto a grate in the furnace. As the fuel falls to the grate it is
partially combusted while in suspension. Combustion of the fuel is
completed on the grate. Figures 77 and 78 are typical spreader-stoker
steam plants currently in operation. This operation can be used with
small plants effecting steam rates of about 25,000 pounds per hour
and with large capacity plants of 500,000 pounds per hour.

Figure 79 illustrates the inclined grate furnace. This
system is very similar to many municipal solid waste incineration
methods. At the furnace inlet the hogged fuel is deposited on the
top section of the grate. The wood fuel then passes through the
three zones of the grate. The first section dries the wood for
combustion in the second section. In the third section the combustion process is completed and the ash is removed.

The York-Shipley Company has a fluidized bed wood waste
heat recovery system that can burn hogged wood residues with moist-

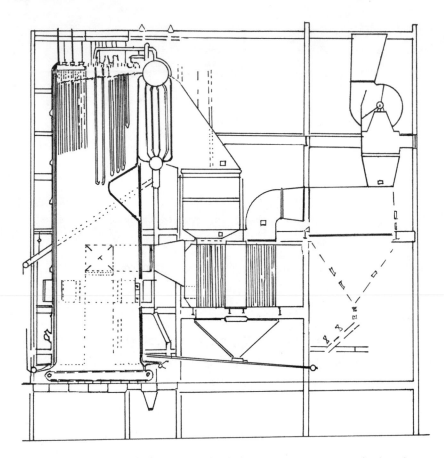

FIG. 77. Steam boiler at a Louisiana paper company, designed to burn hogged bark with a spreader stoker. The plant can generate 450,000 lb of steam per hour.

ure contents of up to 122 percent (dry basis), or 55 percent wet

basis. After the system startup, no supplemental fuel is required

to keep the wood burning and the system is generally automatically

operated. Boiler efficiencies obtained from this fluidized bed

heat recovery unit are comparable to conventional fuel equipment.

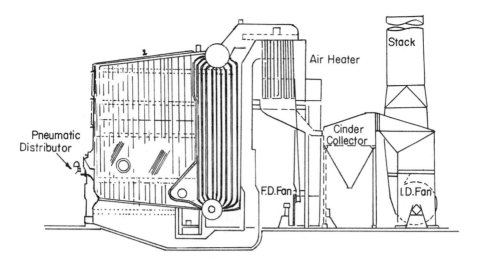

FIG. 78. Recent steam plant installation in Idaho, designed to produce 180,000 lb of steam per hour using hogged fuel. The fuel is spread by pneumatic spreaders.

Figure 80 is a flow sheet of the system's complete operation.

Table 113 examines some of the York-Shipley units displaying their input and output requirements. They have units that range in steam capacity from 3,800 to 60,000 pounds per hour.

Table 114 gives some typical waste wood data. A representative boiler's operating conditions and its requirements for various fuel oils are shown. The typical fuel oils and natural gas are compared to quantities of wood with equivalent energy contents. Also noted is a unit of wood's energy output expressed as various energy output units.

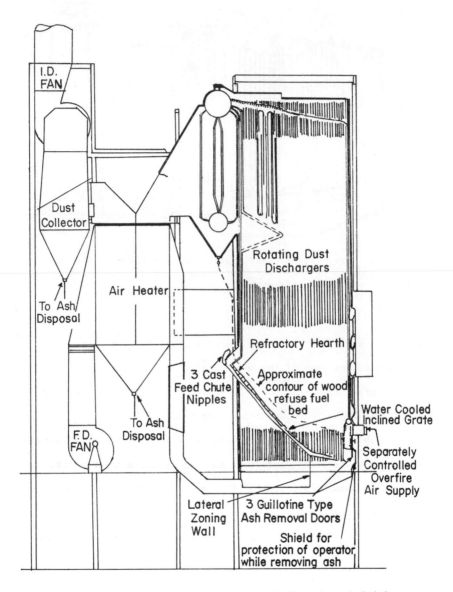

FIG. 79. Inclined, water-cooled grate boiler at a British Columbia pulp and paper company, designed to produce 250,000 lb of steam per hour. The fuel is hogged bark and wood refuse combined with oil or natural gas.

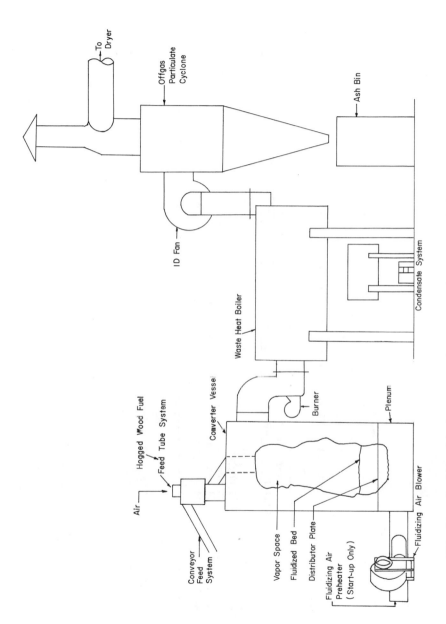

FIG. 80. Fluidized bed waste wood converter flow sheet with waste heat boiler.

TABLE 113 - USABLE STEAM ENERGY FROM HEAT RECOVERY BOILER[132]

| Model No. | Converter Output - Boiler Input-BTUH | BTU/Hr | Maximum Boiler Output | |
			Steam #/Hr	Boiler H.P.
FB-50	5,000,000	3,750,000	3,800	110
FB-75	13,000,000	9,750,000	10,000	290
FB-100	26,000,000	19,500,000	20,000	580
FB-140	52,000,000	39,000,000	40,000	1,160
FB-180	78,000,000	58,500,000	60,000	1,740

TABLE 114 - ENERGY AVAILABLE IN WOODWASTE AND EQUIVALENT CONVENTIONAL
FUEL REQUIREMENTS[128]

WOODWASTE DATA

One - Boiler Horsepower - Input	41,843.75 BTU/Hr.
Output - (Efficiency 80%)	33,475.0 BTU/Hr.
	34.5 LBS/HR STEAM (F/A 212°F)

HEATING VALUE - WOOD

Range -	7,587 to 11,320 BTU/LB
Generage Average -	8,665 BTU/LB

(BASED ON GEN. AVE. - 8,665 BTU/LB)

5 LBS. Wood (4.82)	1 - Boiler HP
1 LB. Wood	7.15 lbs. Steam F/A 212°
1 LB. Wood	6,945 BTU/hr Output
One Unit Wood	200 cu.ft.
One Unit Wood	1900 lb. Dry Douglas Fir
One Unit Wood	16,463,500 BTU Input
One Unit Wood	13,170,738 BTU Output @ 80% Boiler Efficiency
One Unit Wood	393.45 Boiler HP Output @ 80% Efficiency
One Unit Wood	13,574 lbs/hr Steam Output @ 80% Efficiency
1 Gallon No. 2 Fuel Oil 138,000 BTU/Gal.	- 15.92 lbs. Wood @ 8665 BTU
1 Gallon No. 4 Fuel Oil 144,800 BTU/Gal.	- 16.71 lbs. Wood @ 8665 BTU
1 Gallon No. 5 Fuel Oil 146,500 BTU/Gal.	- 16.90 lbs. Wood @ 8665 BTU
1 Gallon No. 6 Fuel Oil 149,000 BTU/Gal.	- 17.19 lbs. Wood @ 8665 BTU
1 Boiler HP Input	.303 GPH No. 2 Oil
"	.288 GPH No. 4 Oil
"	.285 GPH No. 5 Oil
"	.280 GPH No. 6 Oil
"	41.843 cu.ft. Nat. Gas (1000 BTU/cu.ft.)
1 Cu.Ft. Nat. Gas (1000 BTU) -	.1154 lbs. Wood
1 lb. Wood	- 8.665 cu.ft. Nat. Gas

Wood Waste Boiler Performance

The efficiency of a boiler system is dependent on the heat lost during the entire combustion and power generation cycle. The heat balance includes:

1) Heat transferred to the boiler fluid (i.e. steam).

2) Heat lost to gases exiting the stack.

3) Heat lost evaporating moisture in the fuel.

4) Heat lost from the formation of water from hydrogen
in the fuel.

5) Heat lost because of incomplete combustion.

6) Heat lost to radiation.

7) Heat lost to accounting procedures.

Heat losses resulting from conditions (5), (6), and (7)
are small, usually in the range of 4 percent. Heat loss due to
water formation from hydrogen in the fuel is dependent on the burning temperature, and for Douglas fir bark fuel is about 7 or 8 percent.

Because of their high moisture contents, most wood and
bark fuels have high heat losses due to water evaporation from the
fuel. Figure 81 shows the percentage heat loss of Douglas fir

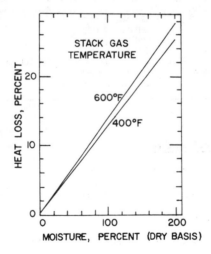

FIG. 81. Heat loss because of moisture content as a percentage
of heating value. The fuel is assumed to be Douglas fir bark with
a heating value of 9,400 BTU per pound.

bark fuel as its moisture content is increased. On the average, a
fuel with a 100 percent moisture content (dry basis) requires about
13 percent of the fuel's total heat output to evaporate the moisture.
Over one-quarter of the final fuel heat output is required at a
moisture content of 200 percent (dry basis). As the moisture
content is increased the flame temperature is lowered and combustion is inhibited, reducing the steam output of the boiler.
The wood's combustion can no longer be self-sustaining as the
moisture content approaches between 180 and 230 percent (dry
basis). Excessively moist wood fuels would probably have to be
dried prior to combustion or be burned along with supplemental
fuels such as oil and coal.

        The amount of excess air required for combustion of the
fuel and the gas exiting temperature determine the boiler heat
loss due to dry stack gases. The heat loss due to dry stack gases
as a function of gas exit temperature and excess air used is shown
in Figure 82 . By reducing the amounts of excess air used, and by
passing the stack gases through a heat recovery unit before they
exit the stack, heat losses can be minimized.

        The overall efficiency of a wood burning boiler system
can be calculated by evaluating the previously mentioned heat losses.
In Figure 83 , a steam plant which burns Douglas fir bark fuel with
40 percent excess air has its overall efficiencies given as a function of fuel moisture content and stack gas temperature. (40 percent
excess air is a normal operating condition for a representative wood
burning plant.) Figure 72 can be used to correct when larger excess

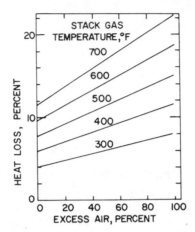

FIG. 82. Heat loss because of dry stack gases as a percentage
of the heating value for Douglas fir bark fuel. Complete combustion
was assumed, and entering air was assumed to be at 70°F.

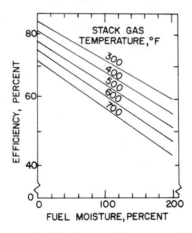

FIG. 83. Calculated efficiency of a steam plant using Douglas
fir bark fuel with 40% excess air. Heat loss from unburned fuel,
radiation, and unaccounted for was assumed to be 4%.

air volumes are needed for complete combustion. A boiler efficiency
of nearly 70 percent can be expected from stack temperatures of
400-500°F and moisture contents of 100 percent (dry basis). These
high efficiencies usually require fuel pretreatment (i.e. predrying)
and stack gas heat utilization via exchangers.

Air Pollution Control

As previously noted, wood and bark fuels have relatively
small amounts of sulfur. Sulfur emissions are usually well below
governmental regulations. Of more concern to wood and bark fueled
power plants are visible plume and particulate matter emission
standards. Oregon's regulations for new boiler units allow 0.1
grain of particulate matter per standard cubic foot of gas. Particulate emissions effected by bark fueled furnaces normally range
from 0.5 to 5.0 grains per standard cubic foot.

The amount and type of particulate matter generated by
wood and bark furnaces is dependent on the fuel burned. Ash
contents vary with wood and bark type. They are generally higher
in bark which also accumulates large quantities of dirt and sand
from handling operations. These emissions are comprised of dirt,
sand and char. The sands and dirt are relatively large particles
and are the nearly invisible component of the emissions. The char,
or unburned carbon, is relatively small and highly visible.

Depending on particle size, power facility size, quantities of exhaust gases, and emission rates, air pollution control
devices to collect these particles can be installed. Larger
particles can be efficiently collected by mechanical cyclones or

a series of cyclones. For smaller particles electrostatic precipitators may be necessary to effect their removal from exiting
gases.

These pollution devices may require large capital and
operating investments. However, depending on the char quality,
the collection equipment can be cost effective. By first passing
the stack gases through a screening process to effect removal of
the larger sand and dirt particles, the char can be reinjected
into the furnace to complete combustion with a minimum increase
in emissions.

Collection devices for wood and bark fueled power plants
usually include a two-stage collection system. However, because
there is relatively little experience with control devices for wood
fired furnaces many collection systems have not wood fuel proven.
Baghouses, wet scrubbers and electrostatic precipitators can be
employed depending upon the plant's requirement and their economic
feasibility at that plant.

Economics

Table 114 shows the wood quantities with heat values
equivalent to the various oils. Table 115 gives the costs for
fuels with equivalent heat contents or per million BTU, and Figure
84 is a comparison of costs of some industrial fuels for steam generation based on data in Table 115. These figures are based on costs
for fuel in December of 1972 and they show hogged wood fuel to be
considerably lower in cost than conventional fuel. Today's oil
prices are shown in Table 116 and probably justify the increased

TABLE 115 – FUEL COST COMPARISON OF SOME INDUSTRIAL FUELS USED FOR STEAM GENERATION

Kind of Fuel	Quantity Of Measure	Cost Per Quantity Dollars	Heating Value Per Quantity Million Btu	Assumed Steam Gen. Efficiency Percent	Fuel Cost $ per million Btu
Oil					
No. 5 Fuel Oil	Barrel	4.80	6.3	80	0.95
No. 6 Fuel Oil	Barrel	4.20	6.3	80	0.83
Natural Gas					
Industrial Firm[1]					
100,000 therms per month	Therm	0.0661	0.1	76	0.87
500,000 therms per month	Therm	0.0612	0.1	76	0.81
Industrial Interruptible[2]					
100,000 therms per month	Therm	0.0480	0.1	76	0.63
500,000 therms per month	Therm	0.0444	0.1	76	0.58
Wood-Bark Residues					
Douglas fir sawdust	Unit[3]	2.0-4.0	16.9[4]	66[5]	0.18-0.36
Western hemlock sawdust	Unit[3]	2.0-4.0	14.3[4]	58[5]	0.24-0.48
Douglas fir hogged bark	Unit[3]	2.0-4.0	24.4[4]	67[5]	0.12-0.24
Western hemlock hogged bark	Unit[3]	2.0-4.0	19.6[4]	66[5]	0.16-0.31

1. Northwest Natural Gas Co., Schedule 21, high load factor, additional charge for excess peak period usage. Effective November 14, 1971.
2. Northwest Natural Gas Co., Schedule 23, Effective November 14, 1971.
3. A unit is 200 cubic feet of bulk volume assumed to contain 1,900 pounds of dry Douglas fir, sawdust, 1,700 of western hemlock sawdust, 2,600 of Douglas fir bark, and 2,200 pounds of dry western hemlock bark.
4. Higher heating values per pound, dry, assumed; Douglas fir sawdust, 8,900; western hemlock sawdust, 8,400; Douglas fir bark, 9,400; and western hemlock bark, 8,900.
5. Efficiencies assumed 40 percent excess air, 500 °F stack temperature, fuel moisture 80 percent for Douglas fir sawdust and bark and western hemlock bark, 120 percent for western hemlock sawdust.

TABLE 116 - VALUE OF AVERAGE TON OF WOOD WASTE USED FOR FUEL[132]

Moisture Conent Of Wood Waste		Average BTU Content		Equivalent #2 Oil		Comparable Value					
		Lb.	Ton			30¢ Oil	35¢ Oil	40¢ Oil	45¢ Oil	50¢ Oil
100% M.C.	(Dry)	4M	8MM	57 Gal.	=	$ 17.10	$ 19.95	$ 22.80	$ 25.65	$ 28.50
50% M.C.	(Wet)									
43% M.C.	(Dry)	6M	12MM	86 Gal.	=	25.80	30.10	34.40	38.70	43.00
30% M.C.	(Wet)									
11% M.C.	(Dry)	8M	16MM	114 Gal.	=	34.20	39.90	45.60	51.30	57.00
10% M.C.	(Wet)									

capital and operating costs of hogged fuel utilization. These
result from the larger weights and volumes of hogged fuel as compared with conventional fuels of equivalent heating content and
therefore, higher transportation and handling costs. Figure 84
illustrates the differences in oil and hogged fuel quantities with
equivalent heating contents.

Steam plants that utilize hogged fuel usually cost more
than a conventional fuel fired plant because of the high moisture
content and handling problems of hogged fuel. Studies have shown
that a boiler system burning hogged fuel alone or with a supplemental
fuel would cost twice as much than a boiler that burns oil only.
However, although the wood and bark fueled plant requires twice the

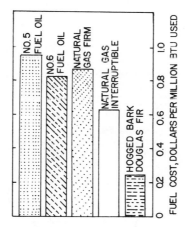

FIG. 84. A comparison of costs of some industrial fuels for
steam generation from data in Table 115. Natural gas usage assumed
was 100,000 therms per month, and the cost of hogged Douglas fir
bark assumed was $4.00 per unit.

initial capital investment over conventionally fueled plants, other

cost effective factors should be considered including:

      1)   Lower hogged fuel costs resulting in long term savings,

      2)   Less air pollution emissions,

      3)   Increasing conventional fuel costs,

      4)   Conventional fuel shortages,

      5)   Use of hogged fuel with supplemental fuel,[133,134,135]

      6)   Wood and bark waste disposal must be effected.

More Wood Derived Waste Disposal and Utilization

     Wood wastes can be utilized in a number of ways as an energy

source. Large power plant installations are not the only means of

energy recovery. Logs for home use are made from sawdust, wood chips,

and other combustibles mixed with wax. Logs made from leaves have

currently become a profitable enterprise. Not only are the leaves

effectively disposed, but their energy is recovered.

     The leaves are separated, dried and shredded before they

are mixed with wax to aid combustion. They are then compressed and

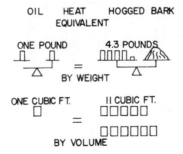

     FIG. 85. Weight and volume comparison of equivalent heat for
oil and hogged Douglas fir bark with 80% moisture.

formed into 16 inch long by 4 inch diameter logs. The photographs

show the plant in Clifton, New Jersey, and the processes the leaves

pass through before final product distribution.[135]

All wood wastes could be treated in a similar manner to

alleviate just a fraction of the energy pinch. These wastes are

available all around us, and are usually obtainable to the fuel

processor at no cost. Sometimes the waste producer will even pay

the fuel processor to remove his wastes. There are large profits

to be made in waste recycling and many individuals have begun to

work at it.

FIG. 86. The basic ingredient to a new industry -- tons of
leaves. This pile represents 30,000 cubic yards gathered by the
City of Clifton, N. J.

FIG. 87. The leaves are separated, dried, shredded, and mixed
with coloring agents and wax.

FIG. 88. The heart of the system, the "Lumber Jack" extruder by the Bonnot Company. Equipped with a 10 H.P. motor it can produce 20 logs a minute and its high-speed guillotine cutter cuts logs to exact length.

FIG. 89. This log being extruded is 4 inches in diameter, 16 inches long, weighing about 6-1/4 pounds, rated to burn 3-4 hours at from 90-115 BTU's.

FIG. 90. Packaging the Ecol-O-Log is the final operation in the
journey that has transformed waste leaves into an energy source that
will delight people all over the country.

Sources of Sludge

Wastewater treatment is utilized to inhibit detrimental
effects to the environment. One of the primary aims of wastewater
treatment is the removal of suspended solids. Suspended solids
are present in wastewaters in varying quantities and in many forms
depending on the origin of the wastewater. During the treatment of
the wastewaters additional solids are generated by the various
biological and chemical processes used to reduce water contamina
tion.

The numerous types of suspended solids found in wastewaters are commonly described by the term "sludge." Sludge in
varying amounts and types are produced by the different methods of
wastewater processing. Table 117 gives the sludge volumes generated by some typical treatment processes.

Wastewater sludge is being generated in enormous quantities and its disposal is of major concern, developing a dilemma:
as the wastewater problem is being resolved adverse environmental

TABLE 117 - TYPICAL SLUDGE VOLUMES PRODUCED BY THE CONVENTIONAL
TREATMENT PROCESSES[137]

Wastewater Treatment Process	Gallons of Sludge Produced/Million Gallons Wastewater Treated			
	Source 1	Source 2	Source 3	Source 4
Primary sedimentation	2,950	3,530	2,440	3,000
Trickling filter	745	530	750	700
Activated sludge	19,400	14,600	18,700	19,400

effects are being created by sludge disposal methods. Figure 91
illustrates the unit processes for sludge processing and disposal.
Figure 92 describes the purposes of these processes and their
effects on the sludge. Sludge processing includes:

Sludge thickening is used to increase the solids
concentration of the sludge and is accomplished with the
following methods:

1) Gravity

2) Flotation

3) Centrifugation

Sludge stabilization is used to remove from the raw
sludge its odor, pathogens, putrescibility rate and other offensive
characteristics with the following processes:

1) Anaerobic digestion

2) Aerobic digestion

3) Lime treatment

4) Chlorine oxidation

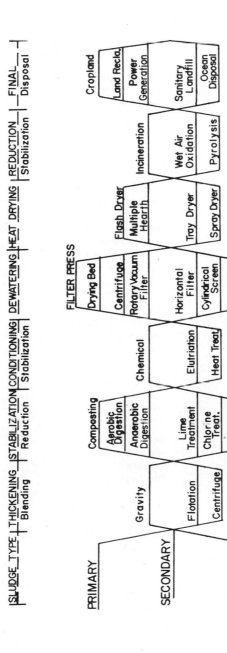

FIG. 91. The unit processes of sludge processing and disposal.[137]

UNIT PROCESSES   FUNCTIONS

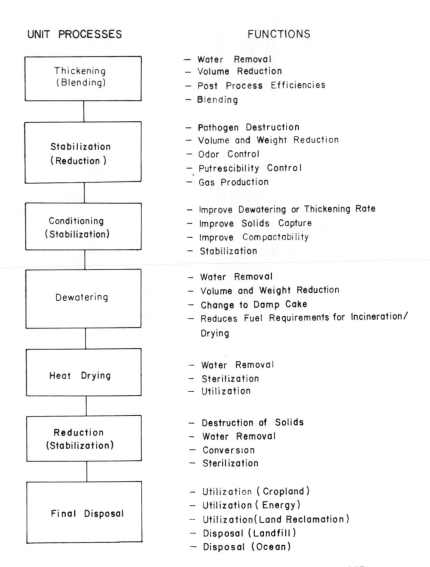

FIG. 92. Sludge processes and their functions.[137]

5) Heat treatment

6) Composting

Sludge conditioning effects water removal with the following methods:

1) Chemical

2) Elutriation

3) Heat treatment

Sludge dewatering converts the slurry sludge into a more solid form with the following:

1) Rotary vacuum filters

2) Centrifuges

3) Drying beds

4) Filter presses

5) Horizontal belt filters

6) Rotating cylindrical devices

7) Lagoons

Sludge is heat dried in dryers such as:

1) Multiple hearth

2) Flash dryers

3) Tray dryers

4) Spray dryers

Sludge reduction to reduce volatile sludge solids utilizes:

1) Incineration

2) Wet air oxidation

3) Pyrolysis

Final sludge disposal methods include:

1) Cropland application

2) Land reclamation

3) Power generation

4) Sanitary landfill

5) Ocean disposal

Thus, sludge processing is costly based on both initial capital
and operating investments. Methods to reduce these costs through
sludge utilization are being examined. Of the seven processes
described above, four of them, stabilization, heat drying, reduction, and final disposal, exhibit energy recovery potentials:
Heat drying and solids reduction through the utilization of
waste heat, final disposal through the burning of the sludge,
and stabilization through the production of methane by the
anaerobic digestion of the sludge.

Anaerobic Degestion of Sludge

Anaerobic digestion as previously defined is a series
of complex biochemical processes in which several groups of anaerobic and facultative organisms decompose organic matter in the
absence of free oxygen. The process is accomplished by gas formation, stabilization, complex structure reductions, pH depression
and release of bound water. This occurs in a mixed environment of
microorganisms where specific groups are active in two distinct
stages -- acidification and gasification. The digestion process is
not complete since organic reduction is approximately 50 percent.

The products of metabolism include methane, carbon dioxide, hydrogen, hydrogen sulfide, organic acids, ammonia, and inert material.[138,139]

The objectives of anaerobic digestion are as follows:

1) Complete decomposition of the organic fraction to a stable form.

2) Sludge volume reduction by decomposition of organics, release of bound water, and thickening.

3) Production of useful byproducts, methane, and a soil conditioner.

4) Reduction of pathogenic organisms.

5) Provide a buffer system for treatment process via sludge storage and reduction.

It must be stated here that operating data indicate the above objectives are often met for primary sludge. However, when more than 50 percent waste activated sludge is combined with the primary sludge, the digested solids concentration can be less than the original concentration. This is particularly true of the high rate system where bound water in the sludge is not released but retained.

First Stage Anaerobic Digestion

Complex reactions involving first stage facultative and anaerobic acid forming bacteria metabolize solid and dissolved complex matter, such as proteins, carbohydrates, fats and other organics, in sewage to form simpler organic structures. These structures are then metabolized through various intermediates producing simple volatile organic acids such as acetic and propionic acids, as shown in Figure 93. In this stage, minor COD and BOD reduction are noted. Some lowering of pH occurs, important buffering materials are produced, and these bacteria have a generation time in hours.

Second Stage Anaerobic Digestion

Simple volatile acids are metabolized by the methane producing bacteria with formation of methane and carbon dioxide as the main end products. Most of the COD and BOD reduction occurs as a result of methane formation. Since the generation

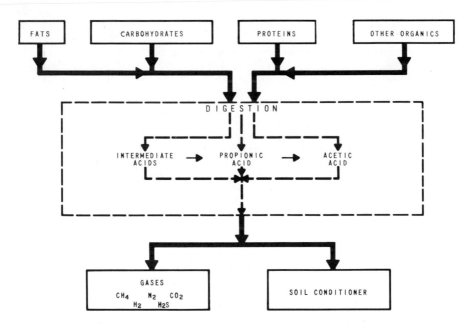

FIG. 93. Typical digestion process.

time for methane bacteria is approximately 2 to 20 days, one can see that this is the limiting or controlling parameter.

During the first and second stage reaction, methane and hydrogen are produced while sulfate and nitrates are reduced. There is a COD or BOD reduction associated with each of the above. Typical values are one cubic foot of methane for 0.14 - 0.18 pounds of COD or BOD reduction.[137,138,139]

Types of Anaerobic Digestion Processes

Presently the standard rate and high rate systems are the two main digestor designs accepted in the U.S. There are four distinct types in active use, two of which are combinations of the two main digestor designs:

    1) Conventional Rate Digestion - One Stage

        a) Heated or unheated

        b) Detention time 30 - 60 days

        c) Solids loading 0.03 - 0.10 lb VSS/cu.ft./day

        d) Intermittent feeding and withdrawal

        e) Process feature - stratification (see Figure 94)

    2) High Rate Digestion - One Stage

        a) Heated to 85° - 95°F (Mesophilic range)

        b) Detention time 15 - 20 days

        c) Solids loading 0.10 to 0.20 lb VSS/cu ft/day

        d) Continuous or time clock feeding and withdrawal

        e) Process feature - homogeneity (see Figure 94)

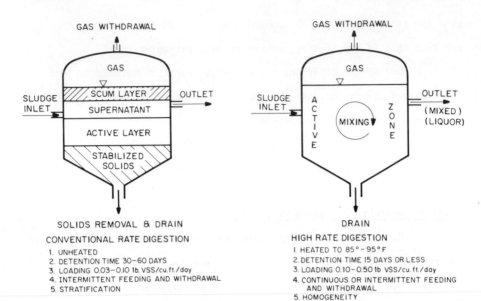

FIG. 94. Digestion systems.

3) Two-Stage Digestion

Combinations of 1 and 2 above (see Figure 95)

4) Anaerobic Contact Process

Similar to process 3 except sludge from the second
stage is returned to the head of the first stage,
similar to the return activated sludge step in an
activated sludge plant. (See Figure 96 showing
with and without thickening.)

The conventional rate, one-stage digestion process, is
shown in Figure   94 . Fresh sludge or combinations of sludges are
usually added on a time clock basis at a controlled feed rate,
say 2, 3 or 4 times a day. The process features stratification,

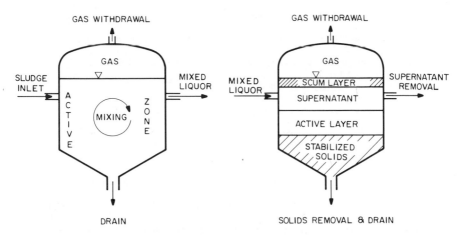

FIG. 95.   Two-stage anaerobic digestion.

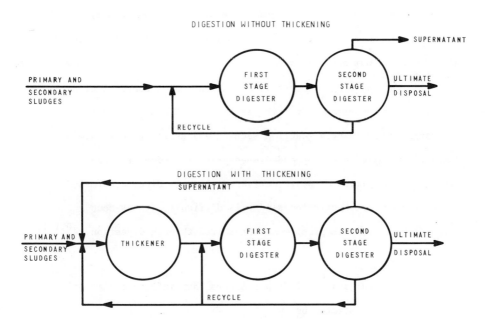

FIG. 96.   Anaerobic contact process.

a gas section, a liquid zone composed of a scum layer on the top, a relatively clear supernatant zone, and active layer where decomposition occurs and a stabilized solids layer at the base. Supernatant is commonly returned to the head of the plant. This practice usually causes problems such as accumulation of fines in the treatment process and return of high dissolved solids such as nitrate for reprocessing.

The high rate, one-stage system process is shown in Figure 94. To enhance this process, post thickening is desirable. This process is successful in upgrading conventional rate because of the mixing aspects. However, this system is incomplete and a further process step is required. Two common solutions are a two-stage system, as shown in Figure 95, or providing a settling tank usually preceded by vacuum degasifier.

The two-stage process is a combination of conventional and high rate processes. It combines the good features of both processes by providing a reaction section and settling section. Operating problems are compounded when primary sludge is mixed with large quantities of waste activated sludge. Since a normal thickening of waste activated sludge is difficult, it has been demonstrated that this problem is not solved through digestion.[140,141,142]

Design Parameters

Physical and chemical parameters that influence anaerobic digestion are illustrated below:

Physical Parameters	Chemical Parameters
Detention Time	pH
Temperature	Alkalinity
Influent Solids Concentration	Volatile Acid
Mixing	Nutrients
Solids Loading	Toxic Materials

It is the combination of the physical parameters that allows a digestor design to function. Bacteria growth is stimulated by temperature; therefore, the stabilization rate increases. Figure 97 shows the interrelationship of temperature and detention time. This is a typical stabilization curve since other physical and chemical parameters would generate similar curves.

Most sources indicate a solids retention time (SRT) of 10 to 15 days. This corresponds to the fact that the generating time for the slowest growing methane bacteria is about 10 days in the mesophilic range. Figure 98 shows the interrelationship of SRT, solids loading and volatile solids concentration. Using a high rate solids loading of 0.10 lb VSS/cu ft/day and an SRT of 10 days, a minimum of 1.5 percent volatile solids or greater is required.[133,136,139]

pH and Alkalinity

pH and alkalinity should be discussed at the same time. Figure 99 illustrates the interrelationship of pH bicarbonate alkalinity and CO_2. Desirable environmental conditions are a pH

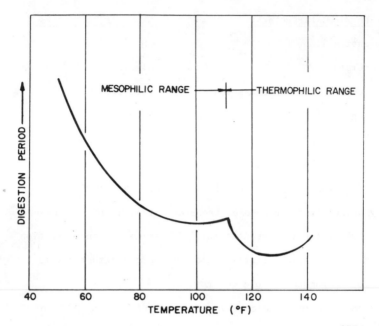

FIG. 97.   Influence of temperature on digestion time.[155]

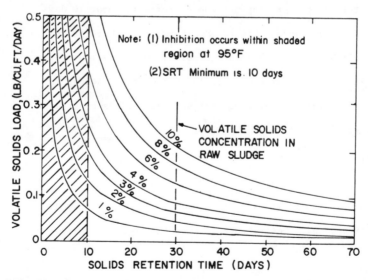

FIG. 98.   Plot volatile solids loading versus SRT for various feed solids.

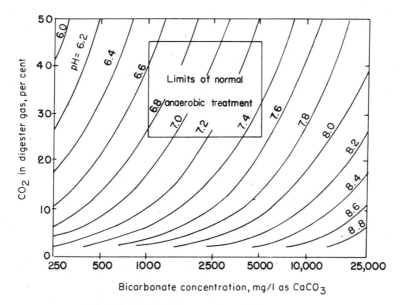

FIG. 99. Relationship between pH, bicarbonate and carbon dioxide concentrations.[142]

of between 6.6 to 7.4, but a tolerable range of 6.4 to 7.8 pH is acceptable.

pH and alkalinity are highly important parameters in anaerobic digestion. The equation dealing with alkalinity formation and dynamics is illustrated here in two forms. For example, the carbohydrate glucose:

I. $C_6H_{12}O_6 \xrightarrow[\text{Bacteria}]{\text{Acid Forming}} 3CH_3COOH$

II. $3CH_3COOH + 3NH_4HCO_3 \xrightarrow{\text{Intermediates}} 3CH_3COONH_4 + 3H_2O + 3CO_2$

III. $3CH_3COONH_4 + 3 H_2O \xrightarrow[\text{Bacteria}]{\text{Methane}} 3CH_4 + 3NH_4HCO_3$

Or, as a system:

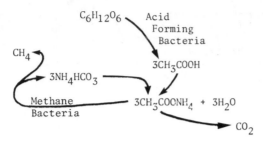

Figure 99 shows a bicarbonate alkalinity range of 1000 - 6000 mg/l as $CaCO_3$. Bicarbonate alkalinity can be determined by laboratory analysis from volatile acids and total alkalinity as shown below:

HCO_3 = Total Alkalinity - 0.8 Volatile Acids

The 0.8 factor converts acid units from mg/l acetic acid to mg/l $CaCO_3$. To maintain good digestion, as shown in Figure 98, the ratio of volatile acids to total alkalinity should be approximately 0.5.[140-147]

Nutrients

Sludges from sewage treatment plants and sludges from meat packing houses contain all the necessary nutrients for good digestion. Wastes containing carbohydrates, fats and other nitrogen and phosphorous deficient wastes require additional nutrient or seed materials.

Toxicity

Toxicity of the wastes can be eliminated or reduced by dilution, precipitation, use of antagonist ions, process by-pass, or process modification.

Studies at various cation concentrations have shown that shock loadings present more problems than gradual toxic increases in concentration. Antagonistic ions can often significantly reduce the effects of cations. With the presence of antagonist ions, the slug to acclimated ratios for common cations such as Na, NH_4, K, Ca, and Mg appear 50 percent higher for the acclimated system. Without antagonist ion, there appears to be a further reduction of 50 percent. Antagonist ion for Na^+ is K^+, for NH_4^+ is Na^+, for K^+ all ions, for Ca^{++} is Na^+ and K^+, and for Mg^{++} is Na^+ and K^+.

Other toxic materials are chromium, copper, nickel, and zinc. Iron and aluminum in low concentration do not appear to be toxic. Sulfides can, on occasion, adversely effect digestor operation. Synthetic detergents and other soluble compounds in excessive concentrations may be toxic. Often a digestor may be subjected to a number of these toxic materials; however, a buffer, antagonist, may already exist in the digester neutralizing any toxic effects.[148-154]

Anaerobic Gas Production

The anaerobic digestion of sludge is a highly complex process that can be drastically effected by minor process operating conditions. The advantages and disadvantages of the system are listed below.[155]

Advantages

1) Elimination of pathogenic organisms.

2) Methane production.

3) High organic loading.

4) No oxygen transfer-aerators.

5) Reduction of fats, carbohydrates, proteins and
cellulose.

6) Reduce volatiles.

7) No additional nutrients required.

8) Low solids production.

9) Useful by-product soil conditioner.

Disadvantages

1) Supernatant treatment.

2) Operation and maintenance problems.

3) Erratic sludge volume reduction.

4) Requires close process control.

5) Upset requires backup systems.

6) Build up of fines in treatment plant.

7) Critical growth system.

8) Oxygen is toxic.

9) External heat required.

10) Tankage expensive.

11) Dewatering.

However, the disadvantages can be effectively offset
through the utilization of sludge gas. Alternate disposal methods
are slowly being classified as environmentally hazardous. The
Marine Protection, Research, and Sanctuaries Act of 1972 and its
Amendments call for the phasing out of most ocean dumping. Landfill space is becoming scarce and as more formally ocean dumped

sludge is landfilled, landfill space will shortly become non-
existent. The enormous volumes of sludge can be utilized as an
energy source.

Sludge gas, for many years, has been used as a fuel
to heat power plant installations. Generally, about 0.2 to 0.3
pounds of solids is generated per capita per day or about 1.0 ton
of sludge is derived from each million gallons of wastewater per
day. Depending on the sludge type and digester process (i.e.
heated or unheated) gas production can vary from 0.32 to 0.74 cu
ft of gas per capita or about 11 to 12 cu ft of gas per pound
of total solids processed. Table 118 gives the characteristics
of some typical sludge gases. The heating value of sludge gas
ranges between 450 and 750 BTU/cu ft depending on the characteristics of the original sludge processed. At a heating value of

TABLE 118 - CHARACTERISTICS OF SLUDGE GAS[137]

Constituent	Values for Various Plants							
	Percent by Volume							
CH_4	42.5	61.0	62.0	67.0	70.0	73.7	75.0	73 - 75
CO_2	47.7	32.8	38.0	30.0	30.0	17.7	22.0	21 - 24
H_2	1.7	3.3	trace	-	-	2.1	0.2	1 - 2
N_2	8.1	2.9	trace	3.0	-	6.5	2.7	1 - 2
H_2S	-	-	0.15	-	0.01-0.02	0.06	0.1	1 - 1.5
H_o BTU/ft^3	459	667	660	624	728	791	716	739 - 750
d_v (air=1)	1.04	0.87	0.92	0.86	0.85	0.74	0.78	0.70- 0.80

about 570 BTU/cu ft, 3.5 cu ft of gas is required to generate
1 kilowatt hour (kWh) of electricity. The heat content of the
gas can be increased by treatment to separate the non-methane
components.

A centralized sludge processing plant to accept the
large volumes of sludge generated regionally could in the future
produce enough methane gas to make electric generation economically
feasible. Furthermore, the upgrading of the sludge off-gas to
pipeline quality could result in metropolitan areas utilizing
sludge gas daily through their natural gas distribution systems.
The costs of anaerobic digestion are relatively high as compared
with the other sludge disposal systems, but this cost can be
reduced through sludge gas utilization.

Sludge the Fuel

Wastewater sludges are often considered low grade fuels,
which is usually the result of their high moisture contents.
Table 119 shows the calorific value of dry raw sludge and the
effects on sludge heating of some treatment methods. These values
are not inconcistent with other waste fuels, however, as noted
in our section on waste wood, the water in a substance contributes
a negative heating value to the entire fuel. Table 120 gives the
representative heating values of some sludge constituents. Thus,
a sludge with a high grease and scum content could be considered
a high grade fuel.

Presently, most sludge reduction processes do not
utilize this heating value of the sludge beyond sustaining its

TABLE 119 - HEATING VALUE OF DRY SLUDGE AND THE EFFECTS OF SLUDGE PROCESSING[137]

Type Sludge	Calorific Value (BTU/lb of Dry Solids)
Raw Primary	9,500
Anaerobically Digested Primary	5,500
Raw (Chem. Precip.) Primary	7,010

TABLE 120 - HEATING VALUES OF SOME SLUDGE CONSTITUENTS[137]

Material	Combustibles (%)	Heating Value (BTU/lb of Dry Solids)
Grease and scum	88	16,700
Raw wastewater solids	74	10,300
Fine screenings	86	9,000
Ground garbage	85	8,200
Digested sludge	60	5,300
Chemical Precipitated solids	57	7,500
Grit	33	4,000

own combustion. Processes such as multiple hearth, fluidized bed and other incineration techniques only utilize the heating value of the sludge for the purpose of combusting it. Usually the moisture content of the influent sludge is high and requires substantial amounts of energy to be burned off. Incineration methods sometimes require supplemental fuels to complete the sludge combustion process.

Heat drying processes offer a viable alternative to the sludge disposal problem. The costs of driving off the

moisture in the sludge by one of the many heat drying processes
can be offset by utilizing the dried sludge for fuel, the
heating value of which can be substantial, as seen in Tables
119 and 120.

In any event, sludge may be environmentally hazardous and must be disposed in a safe manner, the cost of which
is usually considered part of wastewater treatment operating
expenses.   However, as the conventional disposal methods become more and more unavailable the utilization of the heating
value of dried sludge will become a valuable byproduct of the
disposal process both through energy recovery and the net cost
reduction. A typical flash drying system for moisture removal
from sludge is shown in Figure 100.

Pyrolyzing Sludge Fuel

Pyrolysis is a process we have discussed many times
as an energy recovery process for waste products. Presently,
pyrolysis for sludge energy recovery is in its experimental
stages. However, a new process developed by the NASA Jet Pro-
pulsion Laboratory in Pasadina, California, holds the potential
of becoming the solution for sludge disposal and its energy
recovery.[153]

The sludge is pyrolyzed at temperatures of up to 1500°F,
reducing it to a carbon char and small amounts of ash. The carbon
is then steam processed to effect activated carbon, a form of
carbon noted for its high adsorptive characteristics. The activated
carbon is used to effect removal of impurities from fluids extracted

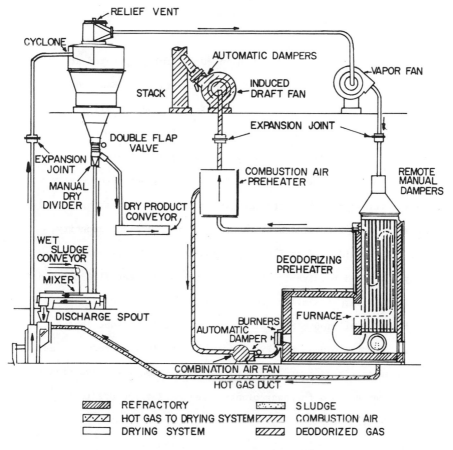

FIG. 100. Flash dryer system.[156]

from the sludge prior to its entrance to the pyrolysis furnace.

The spent activated carbon is then utilized as a fuel in the pyroly-

sis furnace. The pyrolysis gases, similar to pyrolyzed solid waste

gases, are also used as a fuel for the pyrolysis furnace. Sup-

plemental fuel requirements are normally low to negligible. Thus,

the sludge is converted into fuel sources and small amounts of
inert ash.

A small pilot plant has been successfully operated at
Huntington Beach, California, by the NASA Laboratory. Based on
data obtained from this plant, a pilot plant with a one million
gallon a day capacity will soon be constructed at a cost of
$2 million. The USEPA is supplying $1.5 million with the balance
paid for out of state and municipal funds. It is estimated that
costs for this pyrolysis system are about 25 percent less than a
comparable convention sludge tratment facility. A large capacity
sludge plant would be expected to process about 170 million gallons
per day and would require construction costs of between $138 million
and $156 million. A similar conventional sludge plant would cost
more than $200 million. Similar type operating cost reductions are
expected from the sludge pyrolysis system. The fuel byproducts can
be utilized on site and effectively reduce the need for conventional
fuels, or can be sold to power plants and gas distributors to reduce
the net operating cost.

The full energy potential of solid wastes cannot be realized until plans for the total recycling and reuse of the wastes are implemented. Figure 101 shows a recycling plan for industrial wastes recovery. Considerations for the energy potential of solid wastes should include the net energy saved through reprocessing a used material rather than processing the raw material. The recycling and reuse of aluminum results in a 95 percent reduction in energy consumption over the energy required to extract the molten metal from the ore.

Figure 102 shows the flow of solid wastes through a processing plant for recycling purposes. The basis of the entire plant is separation of the solid wastes into its components. Probably the greatest deterrent to recycling has been the inefficiency of the separation processes of a solid waste processing plant. Separation of aluminum, paper, glass and steel from the solid wastes in an efficient manner not only can result in an economically profitable operation, but leads to an easier disposal

437

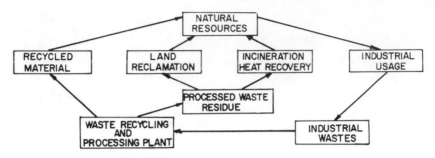

FIG. 101. Flow diagram at industrial wastes recycling.[40]

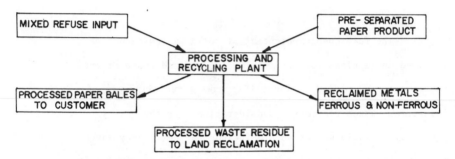

FIG. 102. Solid waste flow through a recycling plant.[40]

process. Many types of separation methods exist to classify the

various components of a solid waste. Some are highly complex

systems and others are quite simple.

Hand Sorting

Probably the simplest of all solid waste separation

techniques is hand sorting. Hand sorting involves people actually

separating specified items from the waste. Every item is hand

picked and routed to either a recycling plant or a disposal process.

The hand sorting process occurs at many stages during

the handling of solid waste. For example, in the disposal of in-

dustrial wastes, reusable items or items which can be sold in bulk
such as scrap metal, wood, paper, glass, etc., are often salvaged
by a hand sorting or picking process. Similarly, construction and
demolition wastes are often hand separated into reusable and disposal
piles. Also municipal or domestic wastes are also sometimes hand
sorted before collection. In some communities, newspapers, glass
and tin cans are collected separately either by alternating collection days (one for garbage type wastes, one for reusable items), or
by industries (soft drink bottlers, aluminum can manufacturers)
seeking a less costly raw material. The occurrence of this separation technique, which is commonly referred to as separation at the
source, is directly dependent upon the existing market conditions
of each particular recycleable material.

The potentials of widespread hand sorting at the source
are subject to many variables. The theoretical efficiency of this
method can compare quite favorably with any other existing method,
but hand sorting at the source relies to a great degree upon the
citizens of a community taking the time and effort to do a satisfactory job. Experiences in many municipalities have indicated
that conscientious separation by citizens usually does not occur
unless some form of compensation is offered. Such compensation can
cause this usually highly economical method of achieving separation
to lose it's economical desirability. As a result, the potentials
of solid waste separation by hand sorting at the source of most
solid waste items is seldom achieved. However, materials such as
newsprint and other paper products are readily reused and can be

economically hand sorted and recycled. Aside from civic and re-
ligious organizations collecting used newspapers to generate revenues, presently two methods are used to collect the newsprint:
1) collection trucks specifically used for newspaper gathering
make regular rounds, and 2) a collection truck with a separate
compartment for newspaper is used to simultaneously collect the
newsprint and other solid wastes. Similarly office and other
commercial paper product residues are sorted and collected.

Another variable to be given extensive study when con-
sideration is given to hand sorting at the source is the relationship between collection of disposals and the markets for possible
reusable materials. Each separated item requires special collection.
Should the market for reusable item be unavailable as to promoting the
economics for private sector collection, then the cost of municipal
collection can cause hand sorting at the source to become economically unfeasible.

In addition to hand sorting at the source, several recycling
plants have operated in the United States and utilized hand sorting
as a separation technique. In 1928 a plant operated in Washington,
D.C. on a continuous conveyor belt hand sorting principle and, in
1945 a more sophisticated plant in Detroit utilized hand sorting
after a series of separation by screens.[158] These old plants have
passed on. However, as a result of hand sorting being an efficient
means of separation, various methods are presently being studied
to determine the most efficient manner to apply this process.
Studies[159] have shown that an efficient hand sorting system would

consist of a picking belt about 24 inches wide for sorters attend-
ing only one side, or 36 to 48 inches wide for sorters on both
sides. Sorters would be spaced about five feet apart and an average
belt speed of 30 to 40 feet per minute should be maintained.   The
solid wastes on the belt should be clean and one unit deep.   In
addition, the entire sorting area should be well lit.

One of the major problems of hand sorting plants is that
the process is undesirable in terms of odor and health standards.
This is especially true for municipal solid wastes separation which
involves garbage and other aesthetically objectionable items.   In
addition, the process involves an initial capital investment and
substantial operating costs.   The cost of labor per ton can be in
excess of $7 or $8 making the process more costly than other disposal
methods.

Magnetic Separation

Magnetic separation is probably the most common mechanical
separation method in use today.   The major portion of the present
magnetic separation technology has been developed by the various
metal industries attempting to reduce the amount of reusable materials discarded.

Magnetic separation basically consists of a conveyor belt
passing refuse past a rotating drum. A stationary magnet is contained within the belt side of the drum. As the drum rotates, ferrous
metals become affixed to the drum and non-magnetic materials pass by
on the belt. The affixed magnetic materials remain on the drum until
the rotation of the drum brings them beyond the stationary magnet.

The ferrous metals then fall off the drum and onto a separate conveyor belt.

There are a variety of different types of magnetic separators which operate on the rotating drum principle. The optimum magnetic separator design requires the length of arc the separated ferrous metal will have to travel along the drum surface be as small as possible to minimize the size of the magnet. Practice has determined that the refuse should then be fed to the magnetic separator at approximately the eight o'clock position when viewing the drum from its end with rotation clockwise. However, use of this design in many cases is impossible because magnetic separators are usually installed as an addition to an existing solid waste processing plant and must be designed to meet a particular plant's requirements. For example, a drum can be installed directly over (bottom feed) an existing belt or can be fed from the top (top feed) where vertical space is limited. Figure 103 shows some typical magnetic drum/conveyor belt arrangements.

The magnetic element should be designed to have a region of metal pickup. The ferrous metals are latched onto and the element's magnetic strength gradually decreases until the materials are released. The most common drums range in diameter from three to eight feet with face widths of up to eight feet.[40]

A typical magnetic separator installation has the magnetic element within the last pulley of conveyor belt. The ferrous metals travel around the pulley to a point under the belt, while non-magnetic materials pass off the end of the belt and are collected separately.

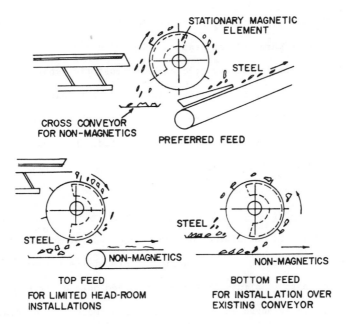

FIG. 103. Various magnetic separator drum arrangements.

A solid waste separation method closely related to magnetic separation is electrostatic separation. Although magnets are not involved in electrostatic separation, the method involves the attraction or repulsion of particles as the driving force of the process. In electrostatic separation, solid wastes are discharged onto a grounded, moving roll and charged by means of an electrode. As a result, the particles which are non-conductors of electricity acquire a charge and are consequently discharged from the roll. The non-conductors are scrapped from the roll to be handled separately.[159] There are many arrangements for electrostatic separation; for example, the separation can be effected by passing the solid wastes between

oppositely charged plates. Depending on the electrical conductivity
properties of each material, separation takes place.[159]

The application of magnetic and electrostatic separation in
the field of solid waste could substantially reduce required disposal
volume and increase recovery and recycling. Although the materials
which are recoverable by these methods are limited, the handling
weight of solid waste and the net operating costs can be significantly reduced by these processes.

Screening

Screening is one of the most common methods of solid waste
separation. Hand sorting is in actuality a form of the screening
separation process. Solid wastes are discharged onto a series of
screens which are arranged in order of decreasing mesh size openings.
The solid waste falls verticaly through the screens and each particle of the waste is collected in accordance to its size. The theory
of screening is based on the fact that generally any solid waste
component has a normal size distribution from which variation is
small. Therefore, separation by size yields an approximate separation by material. This is true to some extent, but screening is
necessary to the separation process.

There are various arrangements of screening processes, including vertical screens, inclined screens, rotary drum screens,
and vibrating screens.[161] In general, screening is a useful tool
for solid waste separation, whose modern procedures provide efficient
and low cost removal of sand, dirt, glass and similar materials from
solid wastes.

Gravity Separation

        More complex systems to effect the separation of the solid
wastes have been developed. These systems are often divided into
broad classes usually determined by the driving force of the methods.
Gravity separation utilizes differences in specific gravities to
classify materials.

        A typical gravity separation process is known as the Dense
Media or Heavy Media method. In this process solid waste materials
are discharged into a prepared solution. The specific gravity of the
solution is adjusted to the specific gravities of the waste materials
desired to be separated. Depending on the waste material's specific
gravity's relation to the solution's specific gravity, the material
will rise or sink. The separated materials can easily be skimmed
off the top or withdrawn from the bottom of the process tank.[162]
Table 121 gives some typical heavy liquids which could be used in
this process. However, the cost of these liquids has limited their
use in large scale operations because they usually cannnot be recycled. The most common heavy medium in operating plants is a

TABLE 121 - TYPICAL HEAVY LIQUIDS USED IN THE HEAVY MEDIA GRAVITY
SEPARATION PROCESS[162]

Heavy Liquid	Specific Gravity	Cost Per Pound (dollars)
Bromoform	2.89	0.530
Ethylene Dibromide	2.17	0.285
Carbon Tetrachloride	1.50	0.108
Trichloroethylene	1.46	0.093

solution of water and fine solids, such as sand, galena, or mag-
netite.[159] The nature of a particular solid waste determines the
specific gravity of the media which is dependent on the specific
gravities of the materials to be separated. In general, dense or
heavy media can have an important role in the process of solid waste
separation, but more research will be required to achieve the maximum
efficiency from the process. Figure 104 shows some typical gravitation
separation processes. Closely related to dense media separation is
another method of gravity separation -- jigging. The jigging process
is also based on specific gravity differences in materials. The
solid waste materials are placed in a perforated container. The
container is continuously submerged and withdrawn from water. The
continuous dunking of the solid wastes results in them becoming
classified according to the differences in their densities. The
materials with the highest specific gravities are on the bottom
and the materials with the smallest specific gravities are on the
top. Jigging has been used as a method of solid waste separation
with some success; however, the major disadvantage of jigging is
that the process consumes large volumes of water which must be
treated at wastewater facilities. Labor requirements for this
process result in high operating costs for this process. Table 122
cites optimum particle sizes for various separation techniques.

The flotation process is considered a method of gravity
separation. Finely shredded solid wastes are discharged into a
liquid filled tank. The flotation process then utilizes air bubbles
to carry selected particles from the solid wastes to the surface

DENSE MEDIA

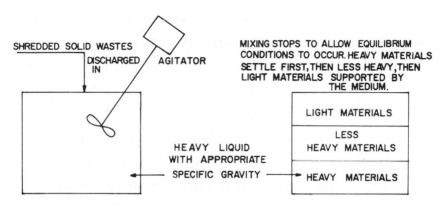

FLOTATION

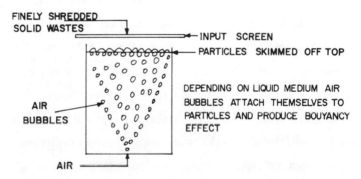

FIG. 104.   Selected gravitation separation processes.

TABLE 122 - OPTIMUM PARTICLE SIZES FOR VARIOUS SEPARATION METHODS[164]

Separation Process	Solid Waste Particle Size (inches)
Magnetic	less than 8
Vibrating Screens	0.125 to 3
Optical Sorting	0.25 to 6
Flotation	0.006 to 0.063

to be skimmed off and stored. Chemicals can be added to a pulp
solution media to make selected waste substances more bouyant or less
buoyant in relation to surrounding particles. Flotation is a highly
efficient process with widespread applications for many waste separation needs. It can usually be applied to any heterogeneous mixture
of particles that are small enough to be lifted by air or gas bubbles.[159]

Probably the most widely used method of gravity separation
is air classification. Air classification basically involves the
separation of solid waste particles by weight. A blower introduces
a jet of air into the solid wastes in a multi-compartment container.
The particles of waste are accelerated both vertically and horizontally. The heavier particles begin to fall into a pre-selected compartment. Lighter particles move on until they fall into their designated
compartments. Air classification is a relatively new method but
as we have noted in our description of current solid waste energy
utilization systems, air classification applications are widespread
and practically unlimited. Furthermore, air classification is
one of the most important processes for the separation of combustible
wastes from non-combustible metals and mineral wastes. It is generally
economical, it doesn't consist of complex equipment and generates no
detrimental pollution effects. Air classification has been successfully
applied in the plastics and food processing industries, and it is an
important step in the recycling of aluminum.

Techniques of gravity separation give great promise of
particle separation of a shredded medium. Continuing research will

refine the processes and lead to a basic application pattern for
solid waste disposal.

Inertia Separation

Inertia separation basically operates on the principle that
various components of solid waste will react differently to applied
forces in accordance with their respective initial momentums. As
a result, inertial separation is an efficient method used to yield
separation of particles by weight. Since the initial momentum of
a particle is directly proportional to its weight, and a similar
force can be applied to each particle, the pull of gravity against
the particle (weight) remains as the major variable governing the
motion of the particle.

There are basically three types of inertial separators:
the Ballistic, Secator, and Inclined Conveyor. Figure 105 gives a
schematic diagram of these three systems. The ballistic separator
utilizes a rotor to accelerate particles horizontally into space.
The particles fall at various distances from the rotor depending
upon their initial momentum. Compartments are provided below the
level of the rotor and collect the fallen refuse. The compartment
closest to the rotor will contain the heaviest materials while
the distant compartments will gather the lighter materials. The
ballistic separator therefore utilizes the force of gravity and
atmospheric friction as the applied forces.

Secator type inertial separators utilize a conveyor belt
to eject the solid wastes against a stationary wall or plate. The

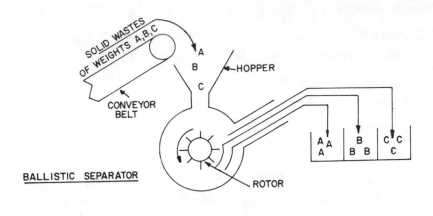

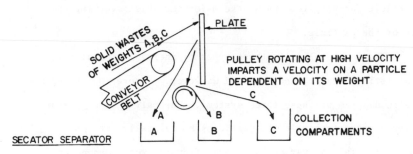

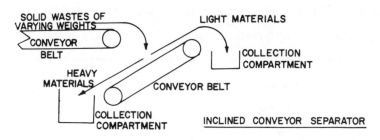

FIG. 105.   Inertial separation equipment.

solid waste then falls onto a pulley which is rotating at high
velocity. Compartments are provided on each side of the pulley.
A particle striking the pulley will either acquire sufficient
momentum to be carried over the pulley and into the distant compartment or will not acquire the required momentum and drop into
the other compartment. In addition, particles which are highly
resilient will be collected in the compartment between the pulley
and the conveyor as a result of bouncing off the stationary wall.
Inclined - conveyor separators are based on a similar principle
as the secator. Solid waste is discharged from one conveyor onto
another conveyor which is inclined to an angle approaching 45 degrees
and rotating at a high velocity in the upwards direction. Compartments are provided at both ends of the inclined conveyor. Lighter
particles landing on the inclined conveyor obtain momentum to be
carried over the top of the conveyor and into the compartment while
heavier particles cannot obtain the required momentum and fall into
the other compartment.

Other Separation Methods

More complex methods of solid waste separation are constantly being developed. Some of the more common methods are optical,
spray drying and the "signature method."

Optical separation is presently used by the glass industries
for the recovery and reclamation of glass. Optical separation is
applied after the glass has been separated from any other refuse
present in the waste. The objective of optical separation is to
separate colored glass from clear glass. Optical separation utilizes

the differences between light transmission properties of clear glass
and the light transmission properties of colored glass to sort the
glasses. The process involves high costs and has not been applied on
a large scale for solid waste glass separation. Figure 106 shows a
typical optical separation operation.

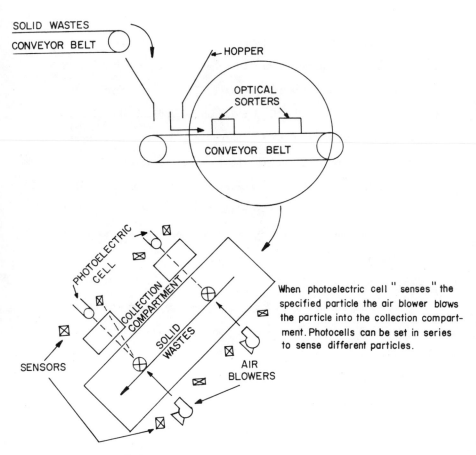

FIG. 106. Optical separation method.

Spray drying is a technique which has been developed to convert a solution to a powder. Spray drying has been utilized by the food and wood processing industries, to economically salvage otherwise non-recoverable solutions or suspensions. In the spray drying process heated air is mixed with a fluid after it has been reduced to a fine spray, to produce a dried powder. The air and powder are then separated. The liquid portion of the mixture is evaporated in the spray dryer and the residual powder is collected and packaged. Spray drying would be used to recover liquid wastes, by condensing them to a powder, and to market the product.

A very sophisticated method of solid waste separation is known as the "signature methods." There are two basic types of signature methods. These are infrared spectroscopy and impact deceleration. The infrared spectroscopy technique uses the diffuse reflection of infrared light which is monitored for characteristic absorption spectra. Light is reflected from solid waste materials in a diffuse form to examine even irregularly shaped objects.

Impact deceleration requires the solid waste materials to be struck by a tool on which an accelerometer is mounted. The principle of both signature separation methods is that these sensors can detect a response induced by either the infrared light or impact tool. The response is recorded and identified by a computer which has stored "signature" responses corresponding to various materials within its memory. The computer controls a separation mechanism·or mechanisms and, depending on the degree of separation required,

the computer orders the separators to remove specified items from a
conveyor belt.

An Operational Separation Process

All the separation methods discussed thus far require the
solid wastes to be shredded prior to processing. The shredding
must be accomplished before any sort of energy or resource recovery
can be instituted. The basic shredding mechanisms include crushers,
cage disintegrators, shears, shredders, grinders, cutters and chippers,
rasp mills, drum pulverizers, disc mills, wet pulpers, and hammermills.

Solid waste size reduction is effected to optimize a system's
efficiency. Both chemical and mechanical separation methods require
a particle size with a reasonable tolerance. The density of
shredded solid waste is more uniform and the mixture is more
homogeneous.

Therefore, the glass recovery process shown in Figure 107
utilizes a shredded solid waste as feedstock. This process, like

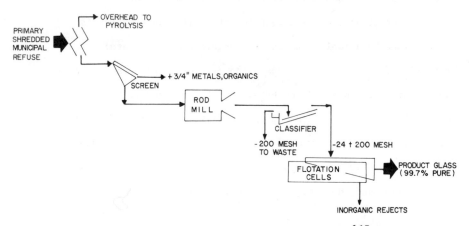

FIG. 107. Garrett glass recovery process.[163]

most solid waste recovery processes, is a combination of the many
separation methods.

The ground solid wastes are first screen separated to remove
large metal and organic particles. They then pass to the rod mill
for size reduction to about + 200 mesh. Depending on the sepearation
process, size reduction varies. Flotation cells require a relatively
fine particle to be transported via air bubbles. The milled wastes
are screened, classified again and then discharged to the flotation
cells. A special glass selective medium fills the cells and as air
bubbles pass through the fluid, glass particles attach themselves
to the bubbles and rise. The non-glass impurities are left behind
to be withdrawn with the spent fluid. The glass product of this
complex system is 99.7 percent pure.

The limiting factor to all solid waste recovery systems is
economics. A useful and viable product must be economically separated
from the solid wastes to make it competitive with virgin products .
The availability of a particular separation system and raw material
does not make resource recovery feasible until a market for the end
product is found.

The energy shortages are not the only problem facing
the industrialized nations as raw materials become short in
supply. Some materials which were readily available can become scarce and others are no longer viewed as having unlimited
supplies. It also takes energy to produce raw materials such as
metals. The concept of supply and demand is fully demonstrated
as shortages have resulted in higher prices for some raw materials.
The higher prices have made it from time to time economically advantageous to recover some of the discarded raw materials which
are in the form of obsolete products or scrap. Waste recovery
has become respectable and economically attractive.

Waste metal interest has also been accentuated by the
huge volumes that are generated as a result of industrial growth
and the rising standard of living. Recovery of the waste materials can reduce the net cost of waste disposal and in some
instances energy requirements are lower for reprocessing than
for processing the virgin ore.

Waste metals exists in all three physical states. Recovery methods for each metal and physical state are different.
Separation of a pure metal from a waste product is hindered
by the difficulties in removing it from other metals in the
product. Waste metals, with the exception of certain scraps, are
not usually found in a pure state. The problem of recovery is
therefore threefold: (1) separation of metals from waste,
(2) separation of one metal from a compound, and (3) purification of the given metal.

The most common methods involve magnetic separation
of the metals into ferrous and non-ferrous. Gravity separation
is used in some instances, as are melting temperature differentials to separate metals. Metals in the liquid state are
separated by reverse osmosis and electrolysis processes.

Solid waste metals, mechanically separated from a
waste stream, usually have paint and other foreign substances
on their surfaces. The removal of these coatings and other
impurities prior to or during metal recovery can generate water
or air pollution problems, and diminish the economic attractiveness of metals recovery.

Sources of Waste Metals

Waste sources originate primarily from agriculture,
urban centers and industry. Waste metal recovery is usually
confined to industrial and municipal wastes.

About 100 pounds of solid waste are generated per
capita and approximately 6 percent are industrial, 89 percent

are mine tailings, smelter slags, dredging spoils, and agricultural wastes, and 5 percent are municipal wastes.

The major portion of waste metals are in the form of
industrial scrap. This scrap is shipped from plants back to
foundries for reprocessing. It is used as an additive to virgin
ores and as the raw material in smelters. Most of the industrial
waste metals are ferrous. Ferrous waste materials amount to 35
million tons annually in the United States including about 8
million automobiles, 21 million large electrical appliances, and
about 50 million large steel containers.

Table 123 gives the quantities of some metals consumed
and the amounts of scrap utilized. For nearly all metals the
consumption rate for virgin and scrap has continuously increased
for the period of time shown. However, the relative usage of
scrap remained constant.

Copper and aluminum are two important metals that are
recovered. Copper is recovered from industrial scrap, machine
parts, transformers, and electrical cables and/or wires. Aluminum
is also recovered from industrial scrap, machine parts and other
discarded industrial products. The recovery of aluminum waste
amounts to about 2 billion pounds or approximately 20 percent of
annual aluminum production. This figure is continually increasing
as more and more aluminum manufacturers turn to less energy de-
manding waste aluminum for processing.

The metal content (on weight basis) of municipal waste
is relatively small. Depending on the source, estimates show

TABLE 123 - CONSUMPTION RATE OF VARIOUS METALS AND THEIR SCRAPS[165]

| Material | Year | | | | | | |
	1963	1964	1965	1966	1967	1968	1969
Aluminum							
Scrap consumption	648	712	817	896	883	1,015	1,057
Aluminum consumption	3,040	3,216	3,736	4,002	4,009	4,662	4,660
Scrap as % of Total	21.3	22.1	21.9	22.4	22.0	21.8	22.7
Copper							
Scrap consumption	1,395	1,553	1,735	1,868	1,541	1,662	1,891
Copper consumption	1,744	1,826	2,005	2,360	1,936	1,863	2,142
Scrap as % of Total	80.0	82.9	86.5	79.2	79.6	89.2	89.5
Zinc							
Scrap consumption	202	226	226	264	263	244	250
Zinc consumption	1,414	1,536	1,742	1,807	1,592	1,728	1,797
Scap as % of Total	14.3	14.7	15.2	14.6	15.3	14.1	13.9
Lead							
Scrap consumption	641	705	748	741	726	726	798
Lead consumption	1,163	1,202	1,241	1,324	1,261	1,329	1,389
Scrap as % of Total	55.1	58.7	60.3	56.0	57.6	54.6	57.4
Tin							
Detinning production	2.3	2.9	2.8	2.6	2.9	2.4	2.4
Secondary consumption	22	24	24	25	23	22	23
Tin consumption	78	82	84	85	81	82	82
Secondary as % of Total	28.2	29.3	28.6	29.4	28.4	26.8	28.0

the average municipal waste to contain about 8-9 percent ferrous metals, 1.0 percent aluminum and 0.4 percent non-ferrous metals. Physical separation methods for metals include hand sorting, magnetic separation, electrical conductivity , radioactivity, jigging and floatation.   Table 124 shows a metal throughput analysis of a typical municipal solid waste. Relatively small amounts of the metal content of municipal solid wastes are re-covered and utilized. The major portion is discarded.

TABLE 124 - METAL THROUGHPUT ANALYSIS OF TYPICAL MUNICIPAL SOLID WASTE[165]

	Input (X 1000 Tons)	Recycled (X 1000 Tons)	Refuse (X 1000 Tons
Ferrous Metals			
Containers and Closures	5,800	400	5,400
Domestic and Commercial Durables	3,800	300	3,500
Total	9,600	700	8,900
Aluminum			
Cans and Closures	310	60	250
Foil	200	--	200
Consumer Durables	200	20	180
Total	710	80	630
Tin			
Tin Plate	31	3	28
Foil	1	--	1
Other	58	21	37
Total	90	24	66
Copper	1,772	1,417	355
Lead	30	--	30

The metal content of solid wastes varies with the
geographical region. Table 125 gives the metal content and a
breakdown as to ferrous, aluminum and non-ferrous content of
the solid wastes of several U.S. locations. One reason for the
variation in aluminum content is the differences in the regional
distribution of aluminum beverage cans.

New York City generates about 30,000 tons of waste per day.
An average metal content of 9.2 percent results in 5,520,000
pounds of waste metal per day or about 1,000,000 tons per year.
Metal wastes in these large quantities can have a very significant

TABLE 125 - METAL CONTENT OF SOLID WASTES FOR VARIOUS U.S. LOCATIONS[165]

City	Metal	Ferrous	Aluminum	Non-Ferrous
Cincinnati	8.7	7.4		
Oceanside, N.Y.	10.6	9.3		
Flint, Michigan	14.5	13.2		
Johnson City	7.5	6.2		
San Diego	7.7	6.4		
Berkeley	8.7	7.4		
Raleigh, N.C.	9.2	8.9		
Santa Clara Co., Calif.	7.4	6.1		
Weber Co., Utah	8.4	7.1		
New Orleans	12.2	10.9		
Alexandria	8.2	6.9		
Tampa		4.8	0.8	0.3
Wilmington		5.7	0.6	0.3
Madison	6.7	5.4		
Perdue	8.0	6.7		
Kaiser	6.9	5.6		
Little	8.7	7.4		
Averages	9.2	7.9	0.9	0.4

monetary value upon recovery. Furthermore, about 80,000 cars are abandoned in New York City and about 85 percent of these cars are recycled. Table 126 gives the prices available for some types of recovered metal wastes and the total annual potential revenues from these wastes. Presently, prices are constantly changing making monetary estimates of metal values extremely difficult.

Methods of Metal Recovery

Resource recovery methods for municipal solid waste is divided into two basic systems. Mechanical separation of materials, which doesn't affect physical properties, is referred to as a "front end system," and about 15 percent of the solid wastes

TABLE 126 - VALUE OF TYPICAL WASTE METALS AND THEIR MONETARY POTENTIAL[165]

	Quantity (X 1000)	Typical Price ($/Ton)	Total Revenue Potential
Ferrous Metals			
Containers and Closures	5,400		
Domestic and Commercial Durables	3,500		
Total	8,900	$15.0	$133,500,000
Aluminum			
Cans and Closures	250.00		
Foil	200.00		
Consumer Durables	180.00		
Total	630.00	$200	$126,000,000
Tin			
Tin Plate	28		
Foil	1		
Other	37		
Total	66	$2,000	$132,000,000
Copper	355	$600	213,000,000
Lead	30	$80	2,400,000
Grand Total			$606,900,000

are separated in this manner. The remaining 85 percent of the solid wastes left after initial mechanical separation is processed in the "back end system." The recovery of metals during these secondary processes is much more difficult and expensive.

Front End Systems

The technology involved in the recovery of materials in their original form (front end systems) is more advanced than back end system technology. The front end system is used to separate solid wastes into homogeneous categories.

Source metal waste separation is the preferred and most efficient method. Most industrial metal processors recover the scrap metal they generate during manufacturing operations. However, domestic source separation is usually negligible except where economic incentives are given, or when other incentives are available to coax people to work a little harder. During World War II the raw material shortages and the high public motivation were factors allowing domestic source separation programs to exist. Affluence and no serous raw material shortages have resulted in the disappearance of these programs. Recently some of these activities are being revived and collection recycling centers have been created. Aluminum cans are one of the major items collected and reprocessed. Collection activities in 37 states collected between 1.3 and 1.4 billion cans in 1972; nearly 70 percent of these cans were in geographical areas where cans are marketed more profusely and local governments aided the campaign.

In Phoenix, Arizona a recycling center was jointly sponsored by the local government and industry in 1971. The Reynolds Aluminum Company and the Arizona Mining Supplies Co. provided the necessary equipment for collection. The city made available land, water, weighing scales, and initial publicity for the project. Other companies have successfully collected and recycled cans.

These solid wastes are separated at the source and hand sorted. Although they are not physically altered during collection, they are not in a pure state and must be processed. Tables 127,128 give a basic analysis of ferrous and aluminum

TABLE 127 - BASIC ANALYSIS OF FERROUS CAN SCRAP[165]

Grade Title	:	Ferrous Can Bundles
Alloy Composition	:	Essentially similar to cold rolled steel used for can stock but containing up to 4 percent aluminum*
Plastics	:	Visibly free except of plastics normally used in can construction
Dirt	:	Less than 1 percent
Loose Organics	:	3.0 percent including small amounts of paint, paper, food, wastes, etc....If incinerated this is reduced to 1.0 percent of burned organics, char and similar residues
Ball Size	:	2 x 2 x 5 feet
Ball Density	:	75 pounds per cubic foot, nominal

*Some 31 percent of all cans produced are tin-free steel and approximately 92 percent of these are bimetal (aluminum ends). The weight of aluminum is 11.2 percent of the can weight. From this data, 3.2 percent aluminum in bundles is a reasonable expectation.

TABLE 128 - BASIC ANALYSIS OF ALUMINUM CAN SCRAP[165]

Alloy Composition	:	Similar to cold rolled steel used for can stock but containing up to 2 percent aluminum
Plastics	:	Usually free except for plastics bonded and riveted to steel
Dirt	:	Less than 1 percent
Loose Organics	:	1.0 percent including small amounts of paint, paper, food, wastes, etc.
Physical Description	:	95 percent minus 1-1/2 inches, balled, loose and free flowing
Bulk Density	:	70 pounds per cubic foot, nominal

cans recovered from scrap. Table 129 gives the basic analysis of other metals recovered.

A typical front end system for metal recovery consists of a receiving area where waste is discharged from trucks. From the receiving area it is transferred to a hammermill or other size reduction unit by large conveyors and cranes. Hand sorting

TABLE 129 - BASIC ANALYSIS OF SCRAP METALS OTHER THAN ALUMINUM AND
FERROUS METALS[165]

Grade Title	:	Other Non-Ferrous Metals
Material	:	Mixed Non-Ferrous Metal
Description	:	Predominantly copper bearing alloys, pieces 2 inches, $\div$ 1/16 inches
Packing	:	Loose, 12-20 pounds per cubic foot
Prohibitive Materials	:	1/2 percent organic, 3 percent dirt, glass
Outthrows	:	Aluminum pieces, magentic ferrous alloys

of the wastes is used prior to the shredder for the removal of

hazardous materials and large metal scraps.

The hammermill and size reduction equipment usually

contain cutters, high-speed crushers, chippers, shredders and

grinders. Air classification of the shredded material is used

almost exclusively to separate the heavy and mostly inert fraction

(non-combustible) from the light, organic, combustible fraction.

Air classification of the waste matter is relatively efficient

and is useful for shredded waste having homogeneous composition.

The heavier elements fall to the bottom of an air classifier and

are withdrawn. The lighter materials are usually collected in

an elevated hopper. This separation method is of little use

to separate the heavy non-combustible material into its components;

therefore, further processing is required.

In the magnetic separation process magnetism is utilized

to remove ferrous metals from the other solid wastes. Low in-

tensity separators of about 5000 gauss are used to separate ferrous

alloys. Materials such as hematite and pyrolusite with weak magnetic properties may be separated from each other with a high in-
tensity field of 20,000 gauss. Belt, induced roll, and drum magnetic separators are commonly used. They usually generqte a magnetic field with permenant magnet because no electrical equipment
installation is required, thus capital and operating plant costs
are reduced. Magnetic separators of this sort are commonly installed in most solid waste disposal facilities.[166]

A relatively inexpensive separation process for magnetic
metals is screening. After passing through the air classifier,
the heavy materials are a conglomeration of many items. In the
screening process items such as glass, ceramic, plastic, rubber,
leather, rags and wood pass through the mesh. The items that are
usually trapped on the screen consist of non-ferrous metals such
as aluminum and other metallic cans.

To further separate the heavy fractions the flotation
separation method can be used. Water flotation can be used as a
two-step process: First to remove organics from metallic materials
and then aluminum from these metals. The liquid medium used
can be varied such that its specific gravity will cause the
separation of a heterogeneous waste into the desired components.
The separation of non-ferrous metals from solid wastes are economically feasible only for large scale waste processing operations
because the non-ferrous metals content of the original waste is
less than 1 percent. A waste processing plant serving a city
with one million inhabitants which generate about 6 million pounds

of waste per day would be left with less than 30 tons of non-ferrous
metals.   Thus, medium and small size solid waste plants that process much smaller amounts of waste usually transport the non-ferrous
metals to secondary metal treatment facilities. At these operations
the non-ferrous metals can be separated into its components with
different degrees of purity.

Back End Systems

        The back end system consists of a number of processes not
only concerned with material recovery, but also with heat recovery,
soil enrichment and the development of high protein feed for cattle
and pigs. Some metals not recovered in a front end system process
can be extracted from incinerator residue. Metals in solution can
be removed and recycled through the utilization of electroplating,
precipitation, and crystallization, centrifuging and reverse osmosis
processes. Table 130 gives the metals content of typical incinerator
residue. In some instances, the metals content of the residue is
greater than the virgin metal in their natural ore.

        Figures 108, 109, 110 show flow schematics of typical waste
processing plants; process for metal contaminant removal from wastewater; and metal reclamation from smelting operations.

TABLE 130 - THE METAL CONTENT OF TYPICAL INCINERATOR RESIDUE[165]

Residue	Percent
Wire and large iron	3.0
Tin cans	13.6
Small ferrous metal	13.9
Non-Ferrous metal	2.8

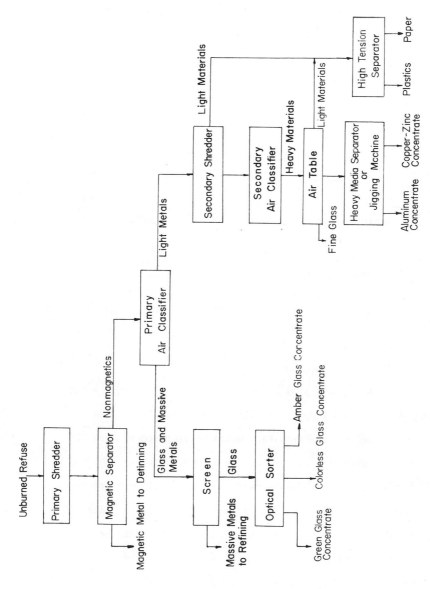

FIG. 108. Typical waste processing plant.

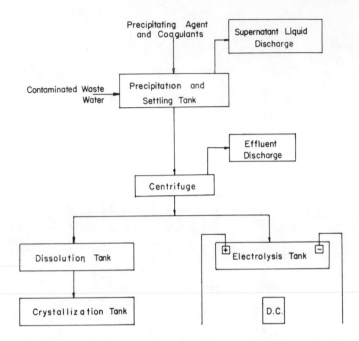

FIG. 109. Process for removing metal compound contaminants from waste water.

Liquid Waste

     Metal recovery from industrial and municipal wastewater in the past involved simple neutralization in acidic or alkaline solutions. However, the resultant metallic compound sludges did not settle or filter easily creating a sludge disposal problem. The water fraction of the sludge was very high, nearly 98 percent. The sludge had to be settled and thickened to reduce the water content and then hauled to landfill sites. However, the metallic content of this sludge was detrimental to the environment, as problems developed from high water tables and leachate causing

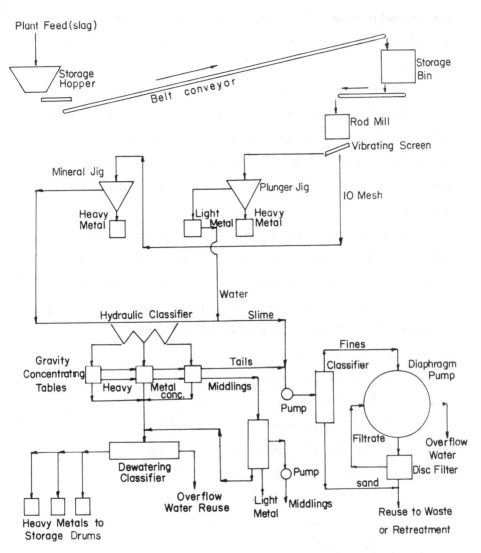

FIG. 110.   Scrap metal reclamation from smelting and electric
furnaces.

water pollution. These effects were long lasting because of the
non-degradable nature of the sludge. Costs of this operation
including several processes and transportation and labor are high
and cannot be economically justified.

A process to separate the metal contaminants from industrial wastewater utilizes chemical additives to precipitate metallic
compounds. Most of the problems of conventional treatment operations
are eliminated with the added benefit of metal recovery for recycle.
One or more chemical agents are added to the metal compound containing wastewater to precipitate out the metal compound, producing an
aqueous metal compound sludge. After dewatering, the aqueous metal
compound sludge is separated from the supernatant liquid and transferred to a centrifuge. The centrifugation process concentrates
the aqueous metal compound sludge. The sludge is then processed
to remove the metal compounds by means of electrolysis, crystallization or other methods.

These methods may include:

1) Ion flotation

2) Carbon adsorption

3) Liquid-liquid extraction

4) Liquid ion exchange

5) Electrodialysis

6) Reverse osmosis

7) Chelation

Recovery of Ferrous Metals

The major portion of the metals in solid wastes is ferrous, or iron containing materials. Industrial waste metals are generally ferrous byproducts from manufacturing and processing operations. The ferrous metal content of municipal solid waste is about 8 percent. Junked automobiles and major electrical appliances are a major source of ferrous scrap. Of the 35 million tons of ferrous metal wastes generated annually, less than half is recycled. Table 131 gives the consumption of ferrous or magnetic metals wastes in various steelmaking processes. All processes show a wide variation in steel production rate over the years analyzed, but the fraction of magnetic metal scrap utilized remains constant.

Industrial metal wastes make up nearly a fifth of the ferrous scrap consumed in steelmaking. These wastes consist of steel byproducts from punch presses, metal working tools, and other manufacturing machines and mills.

The recycling of these waste metals is common practice and economical since they are relatively pure. These wastes account for about 75 percent of the ferrous scrap charged to the steel furnaces. Only about 25 percent of the total ferrous metal waste used by the steel industry comes from landfills, recycling project centers and municipal solid wastes. Junked automobiles are either sheared, broken, compressed, or shredded. The bodies are then compressed and flattened into slabs.

TABLE 131 - U.S. CONSUMPTION OF MAGNETIC METAL WASTES IN VARIOUS STEELMAKING PROCESSES[165]

	Year											
	1960	1961	1962	1963	1964	1965	1966	1967	1968	1969	1970	1971
Basic Oxygen												
% steel production	3.3	4.0	5.6	7.8	12.1	17.4	25.3	32.7	37.1	42.7	48.1	53.1
% scrap of charge	28.1	27.9	26.9	28.0	30.6	29.6	29.0	29.4	29.1	29.9	30.0	----
Open Hearth												
% steel production	87.0	86.2	84.3	81.3	77.2	71.6	63.4	55.7	50.1	43.2	36.5	29.5
% scrap of charge	41.7	40.9	40.3	41.0	41.2	41.6	41.5	41.6	44.1	45.1	45.0	----
Electric												
% steel production	8.4	8.8	9.2	10.0	9.9	10.5	11.0	11.6	12.8	14.1	15.4	17.4
% scrap of charge	96.4	97.3	97.8	98.0	97.9	99.7	98.4	98.0	97.4	98.6	99.0	----
Total												
% steel production	100	100	100	100	100	100	100	100	100	100	100	100
% scrap of charge	45.3	45.0	44.5	45.3	45.2	45.5	44.4	44.2	45.4	46.2	46.0	----

In the automobile process the compressed body is sliced into pieces with a guillotine shear and then shredded into fist-sized pieces. Magnetic separators remove the ferrous material from the rest of the automobile waste. The quality of this ferrous material is nearly equivalent to the original steel and/or iron.

Another operation passes the shredded automobile through a liquid nitrogen freezing process. The metals are brought down to temperatures approaching minus 310°F where the ferrous metals become brittle and the non-ferrous metals become elastic. Further sized reduction processing splinters the ferrous materials like glass while the non-ferrous metals remain intact. An electromagnet easily separates the ferrous and non-ferrous metals. The copper content of the recovered ferrous material is reduced to about 0.05 percent.[167]

The copper content of the recovered ferrous metal is of primary importance to the steel industry. Recovered ferrous metals have to be of high quality and free of impurities, or they have to be processed to effect such a condition. Copper wiring is a major problem as excessive quantities of copper in steel produces faults and surface defects. The copper content of most ferrous metal wastes is limited to 0.06 percent. As a result, auto bodies which contain relatively large amounts of copper parts and wiring must be stripped before processing.

Ferrous wastes have effected the steel industry use of the electric furnace whose feedstock can be composed almost ex-

clusively of waste metal. Electric furnaces in 1960 were responsible for only 8.4 percent of the strel produced. This figure was
more than doubled to 17.4 percent, 20.9 million tons, by 1971.
Electric furnaces utilizing waste material are commonly used in
small mills, having capacities in the 50,000 to 250,000 tons of
steel per year range. The advantage of these small mills is their
ability to be built near the site of waste generation resulting in
low transportation costs both for new feedstock and for marketing
the end product. As waste metals' generation increases so has the
use of the electric furnace and it is anticipated that electric
frunaces will process 30 percent of the 1980 steel production and
50 percent in 1990.

The Basic Oxygen Furnace (BOF) for steel processing has
also effected larger amounts of waste ferrous metals utilized.
The waste metals are preheated to about 1400°F before entering the
BOF. The amount of heat needed in the furnace to attain the normal
operating condition temperature of 2800°F is significantly reduced.
The smaller heat requirement increases furnace capacity by about
40 percent while reducing energy requirements.

The increased utilization of ferrous waste metals has
other advantages. The energy requirements of reprocessing scrap
steel are less than the energy required in producing steel from
iron ore. The energy cost of producing a pound of steel from waste
ferrous metals is equivalent to 0.22 pounds of coal. The energy
required to produce a pound of steel from iron ore is equivalent to
1.11 pounds of coal, or about 5 times as much energy. By reprocessing

waste metals, energy is being conserved that can be utilized elsewhere.
Since about 80 percent less coal is burned, there is a large reduction
in air pollutant emission rates.

Production of steel from ferrous wastes results in
lower particulate emissions than the normal steelmaking process.
Furthermore, electric furnace processing waste metals and not
equipped with air pollution control devices, emit about 11 pounds
of particles per ton of steel produced. Under the same operating
conditions, an oxygen furnace processing iron ore emits about
46 pounds of particles per tone of steel produced. The power
requirements for an electric furnace processing 2000 pounds of
steel are 600 kilowatt hours. In order to generate 600 kWh of
electricity a typical power plant must burn about 400 pounds of
coal with an ash content of 11 percent. The potential particulate
emissions for the processing of a ton of steel are therefore about
100 pounds. These quantities of particulates are not necessarily
emitted, but are controlled only after large capital investments
in pollution control equipment are made usually requiring large
operating expenses. Depending on the initial processing demands
of the iron ore, these pollution levels can be greatly increased.

Aluminum Recovery

The recycling of aluminum is advantageous both economically
and from an ecological standpoint. Although aluminum is in abundance (about 8 percent of the earth's crust), most of the high-grade
natural bauxite deposits lie outside the boundaries of the U.S.
The processing of the lower grade ores results in large energy

requirements. These factors have made the recovery and reuse of
aluminum wastes very attractive. Additionally, practically all
forms of aluminum can be recycled, and this reprocessing has an
energy demand of only about 5 percent the energy needs to produce
aluminum from ore.

The quantities of aluminum wastes have been continually
increasing with increased aluminum usage. Aluminum, a non-ferrous
metal, does not exhibit any unique characteristics that would allow
for its economical recovery through conventional solid waste
separation methods. Recovered aluminum must be of high quality
for reuse, but the separation method utilized cannot be complex
to the point that the recyclable process becomes economically
unattractive.

Figure 111 illustrates a dry separation system for
aluminum. Solid wastes which have been shredded, screened and

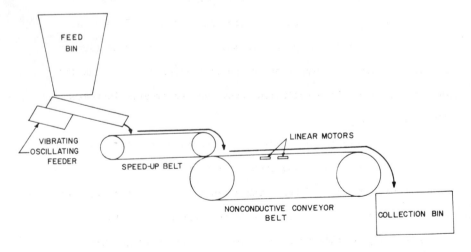

FIG. 111. Garrett "Recyc-al" aluminum separator process.[163]

classified for the removal of combustibles and ferrous metals
are discharged to the aluminum recovery feed bin. Wastes with
aluminum contents usually in the range of 7 to 20 percent feed
onto a speedup belt and then onto a nonconductive conveyor belt.
Linear induction motors are situated beneath this belt. These
motors generate a magnetic field which induces eddy currents in
non-ferrous metals. The aluminum wastes passing over these motors
are effected with a magnetic field of opposite polarity to the
linear induction motor's magnetic field. The opposite fields
repel each other forcing the aluminum scraps to the edge of the
conveyor belt where they are guided by baffles into a storage bin.
This process can recover about 60 percent of the aluminum present
in the original solid wastes.

REFERENCES

1. Kearly, James A. "Long-Range Fuel Supplies, Refuse In Lieu of Natural Fuels." In Solid Waste: A Source of Energy, Proceedings of a two-day program on central heating, The University of Tennessee, Nashville, Tenn., Oct. 11-12, 1973.

2. Kasper, William C. "Solid Waste, Its Potential As A Utility Fuel." Aware, 45, pp. 7-10, June, 1974.

3. Let's Talk Trash, A publication of the National Center for Resource Recovery, Washington, D. C., 1971, pp. 1-8.

4. Solid Waste Management, A Comprehensive Assessment of Solid Waste Problems, Practices, and Needs, Office of Science and Technology, Washington, D. C., May, 1969.

5. Cross, Frank L., Jr. Handbook On Incineration, Technomic Publishers, Westport, Conn., 1972.

6. Anderson, Larry L. "Energy Potential From Organic Wastes: A Review of the Quantities and Sources," U. S. Bureau of Mines, Washington, D. C., 1972, pp. 1-14.

7. From Refuse To Energy: A Proven Solution, Wheelabrator-Frye, Inc., pp. 4-8, 1974.

8. "Town Will Buy Back Its Garbage As Power," Engineering News-Record, vol. 195, no. 5, July 31, 1975 p. 11.

9. Cohan, L. J. and Ferrandez, J. H. Burning Of Waste As Fuels In Industrial Boilers, presented at the APCA annual meeting, Chicago, June 24-28, 1973.

10. Wisely, F. E. City of St. Louis Electric Company Energy Recovery Process Solid Waste As A Boiler Fuel, presented at the proceedings of the annual ASCE meeting on national environment, New York, 1973.

11. Kirshner, Marvin, Feasibility of Refuse Fuel For District Heating, presented at the annual meeting for International District Heating Association, June 24, 1971.

12. MacAdam, W. K. The Economics Of Energy Recovery From Refuse, Wheelabrator-Frye, Inc., New York, Jan. 1974.

13. Fernandez, J. H. and Shenk, R. C. The Place of Incineration in Resource Recovery of Solid Waste, presented at the Sixth ASCE National Incinerator Conference, Miami, May 12-15, 1974.

14. Murray, D. L.; Hartz, K. E.; and Carlson, R. G. The Economic and Technology of Refuse As an Energy Source, presented at the 36th Annual Meeting American Power Conference, Chicago, May 1, 1974.

15. Cost Comparison of Two Methods for Reducing Electric Utility Plant Fuel Oil Requirements by Augmentation with Energy from Refuse, Wheelabrator-Frye, Inc., New York, March 15, 1974.

16. Lowe, Robert A. Energy Recovery From Waste, U. S. Environmental Protection Agency, Washington, D. C., 1973.

17. Dreifke, Gerald E.; Klumb, David L.; Smith, Jerrel D. "Solid Waste As a Utility Fuel," Proceedings of the American Power Conference, vol. 35, 1973, pp. 1198-1206.

18. Skinner, John H. The Demonstration of Systems for Recovering Materials and Energy from Solid Wastes, U. S. Environmental Protection Agency, Washington, D. C., 1974.

19. "ECO-FUEL^{TM} II Process," Combustion Equipment Associates, Inc., and Arthur D. Little, Inc., 1974.

20. "Liabilities Into Assets," Environmental Science and Technology, vol. 8, no. 3, 1974, pp. 210-211.

21. Combustion Equipment Associates, Inc., Annual Report, 1974.

22. Beningson, R. M. "Resource Recovery Revelations," Resource Recovery, April-June, 1974.

23. Sussman, D. B. Baltimore Demonstrates Gas Pyrolysis, U. S. Environmental Protection Agency, Washington, D. C., 1975.

24. Linaweaver, F. P. and Crooks, C. W., Jr. "Pyrolysis for Baltimore: A Dramatic Breakthrough in Solid Waste Management," District Heating, vol. 59, no. 3, 1974, pp. 12-14.

25. Shulz, H. W. Thermal Oxidation Processes, Columbia University, New York, April 15, 1973.

26. Butler, P. "Solid Waste Disposal by Pyrolysis Yields Only Landfill," Process Engineering, March, 1974, p. 6.

27. "Disposing of Solid Wastes by Pyrolysis," Environmental Science and Technology, vol. 9, no. 2, 1975, pp. 98-99.

28. McAdam, W. K. Design and Pollution Control Features of the Saugus, Massachusetts Steam Generating Refuse-Energy Plant, presented at the 67th APCA annual meeting, Denver, June 9-13, 1974.

29. Standrod, S. E., Jr. The RESCO North Shore Facility presented at the U. S./Japan Energy Conservation Seminar, Feb. 4, 1974.

30. Mullen, J. F. and Singer, J. G. European-Closing the Refuse Cycle, presented at the Joint Power Generation Conference, New Orleans, Sept. 16-19, 1973.

31. Astrom, L.; Kranebitter, F.; Strandell, O.; and Harris, D. W. A Comparative Study of European and North American Steam Producing Incinerators, presented at the National Incinerator Conference, Miami, May 12-15, 1974.

32. "Netherlands" (news release), Consulate General of the Netherlands, Commercial Division, New York, July 20, 1973.

33. Kimura, H. and Maekawa, K. "Present Situation of Refuse Incineration in Japan and Problems in Utilization of its Waste Heat for District Heating," Japanese District Heating, June, 1974.

34. Asukata, R. and Kitami, S. Present Situation and Future Trends of Japanese Refuse Incineration Plants with Power Generation, presented at the 1974 National Incinerator Conference, Miami, May 12-15, 1974.

35. Aubin, H. The New Quebec Metro Incinerator, presented at the Sixth ASCE National Incinerator Conference, Miami, May 12-15, 1974.

36. Canadian Steam Generating Incineration Plant for the Urban Community of Quebec, presented at the 66th annual APCA meeting, Chicago, June 24-28, 1973.

37. Ricci, L. J. "Garbage Routes to Methane," Chemical Engineering, vol. 81, May 27, 1974, pp. 58-66.

38. Levy, S. J. Markets and Technology for Recovering Energy from Solid Waste, USEPA, Washington, D. C., 1974.

39. Appell, H. R.; Fu, Y. C.; Friedman, F.; Yavonski, P. M.; and Wender, I. "Converting Organic Waste to Oil: A Replenishable Energy Source," U. S. Bureau of Mines Report of Investigation 7560, 1971.

40. Cheremisinoff, P. N. and Young, R. A. Pollution Engineering Practice Handbook, Ann Arbor Science Publishers, Inc., Ann Arbor, Michigan, 1975.

41. Levy, S. J. "Pyrolysis of Municipal Solid Waste," Waste Age, vol. 6, no. 7, 1974.

42. Drobny, N. C.; Hull, H. E.; and Testin, R. F. Recovery and Utilization of Municipal Solid Waste, USEPA, Washington, D. C., 1971.

43. Second Report to Congress: Resource Recovery and Source Reduction, USEPA, Washington, D. C., 1974.

44. Resource Recovery: The State of Technology, The Council on Environmental Quality, Midwest Research Institute, Kansas City, Mo., Feb. 1973.

45. Landgard Technology, Monsanto Research Corp., Published Information.

46. Purox System Process Description, Union Carbide Corp., Published Information.

47. "Methane From Waste...How Much Power Can It Supply?," Popular Science, vol. 250, no. 6, 1974.

48. Hicks, Gerald Private Correspondence, 1975.

49. Zinn, R. E. and Niessen, W. R. Commercial Incinerator Design Criteria, proceedings 1968 National Incinerator Conference (ASME), pp. 337-359.

50. Warner, A. J.; Parker, C. H.; and Baum, B. Solid Waste Management of Plastics, DeBell & Richardson, Inc., Enfield, Conn., Project 1440.2, 1970.

51. Schonbern, H. The Situation of Waste in Germany and the Plastics, EETA Plastics Association, Oslo, May 26, 1970.

52. Potts, J. E. Continuous Pyrolysis of Plastic Wastes, presented at the National Industrial Solid Wastes Management Conference, University of Houston, Texas, March 25, 1970.

53. Tabakin, R. B. Private correspondence, 1975.

54. Weinstein, N. J. Waste Oil Recycling and Disposal, EPA Report No. 670/2-74-052, Cincinnati, Ohio, 1974.

55. American Petroleum Institute Committee on Disposal of Waste Products, Waste Oil Roundup...No. 3, API Publication No. 1587, Washington, D. C., 1974.

56. Lederman, P. B. and Weinstein, N. J. Waste Oil Management: A Status Review, presented at the API - Automotive Industry Forum, Detroit, 1974.

57. Martin, E. J. and Gumtz, G. D. State of Maryland Waste Oil Recovery and Reuse Program, EPA Report No. 670/2-74-013, Washington, D. C., 1974.

58. Response Analysis Corporation, The Study of Waste Oil Recycling, pending release by EPA, 1974.

59. Arthur D. Little, Inc., Study of Waste Oil Disposal in Massachusetts,
     Commonwealth of Massachusetts, 1969.

60. Report on Waste Oil: Oil - It Never Wears Out, It Just Gets Dirty,
     New York State Senate Task Force on Critical Problems, Albany, 1974.

61. Report to Congress: Waste Oil Study, U. S. Environmental Protection
     Agency, Washington, D. C., 1974.

62. Freestone, F. J. Runoff of Oils from Rural Roads Treated to Suppress
     Dust, EPA Report No. R2-054, Edison, N. J., 1972.

63. Petroleum Analytical Research Corporation, Study on the Use of Waste
     Lube Oil as a Fuel, available thru API, 1973.

64. Information Transfer, Inc., Proceedings of the International Con-
     ference on Waste Oil Recovery and Reuse, Washington, D. C., Feb. 12-14,
     1974.

65. Schroeder, H. "Metallic Menaces in the Environment," Fortune, vol. 81,
     Jan. 1975.

66. Sullivan, J. Marine Pollution by Carcinogenic Hydrocarbons, National
     Bureau of Standards Symposium on Marine Pollution Monitoring,
     Washington, D. C. 1974.

67. Preliminary Report to Congress: Waste Oil Study, USEPA, Washington,
     D. C., 1973.

68. Department of Defense, Waste Oil Recycling Study, Defense Supply
     Agency, Washington, D. C., 1972.

69. Environmental Protection Agency, Lead and Air Pollution: A Bibliography
     with Abstracts, Report No. 45011-74-001, Washington, D. C., 1974.

70. Environmental Protection Agency, The Impact of Oily Materials on
     Activated Sludge Systems, Report No. 12050 DSH, Washington, D. C., 1971.

71. Environmental Quality Systems, Inc., Waste Oil Recovery Practices;
     State of the Art, EPA Report, 1972.

72. Aberdeen Proving Ground, Waste Oil Utilization, Program, 1973.

73. GCA Technology, Waste Automotive Lubricating Oil Reuse as a Fuel,
     EPA Report No. 600/5-74-032, Washington, D. C., 1974.

74. "Burning Waste Oil," Compressed Air Magazine, pp. 14-16, Nov. 1974.

75. Lederman, P. B. Waste Oil Management from Sump to End Use - A Status
     Report, presented at the 1975 SAE Automotive Engineering Congress,
     Detroit, 1975.

76. National Oil Recovery Corporation, Conversion of Crankcase Waste
     Oil Into Useful Products, EPA Report No. 15080 DBO, Washington,
     D. C., 1971.

77. Oil: It Never Wears Out, Just Gets Dirty, Environmental Science
     & Technology, vol. 8, no. 4, 1974, pp. 310-311.

78. Whisman, M. L.; Goetzinger, J. W.; and Cotton, F. D. Waste
     Lubricating Oil Research: A Comparison of Bench-Test Properties
     of Re-refined and Virgin Lubricating Oils, U. S. Bureau of Mines
     Report 7973, Bartlesville, Oklahoma, 1974.

79. Skinner, D. J. Preliminary Review of Used Lubricating Oils in
     Canada, Environment Canada Report No. EPS-3-WP-74-4, 1974.

80. Bonnifay, P. A New Process for Reclaiming Spent Lubricating Oils,
     Institut Francais du Petrole, presented at the National Fuels and
     Lubricants Mtg., New York, 1972.

81. Putscher, R. E. Study of Re-refining Waste Disposal, Armour Research
     Foundation, Lyons, Ill., 1960.

82. Bethea, S. R.; Bosniack, D. S.; Claybaugh, B. E.; and Mohundro, E. L.
     "To Hydrotreat Waste Lube Oil," Hydrocarbon Processing pp. 134-136,
     Sept. 1973.

83. Young, R. A. and Cheremisinoff, P. N. "Waste Oil Reclamation Provides
     New Fuel Source," Pollution Engineering, vol. 7, no. 2, pp. 44-45,
     1975.

84. "Modern Methods for Acid Sludge Disposal," Oil in Canada, pp. 24-27,
     Jan. 18, 1954.

85. Putscher, R. E. Separation and Characterization of Acid Sludge,
     Armour Research Institute, Lyons, Ill., 1962.

86. Leonard, R. P. Brief Investigations on the Treatment and Recovery of
     Resources from Waste Oil Sludges, Claspan No. VT-3044-M-1, Buffalo,
     N. Y., 1973.

87. Twomey, D. W. "Lube Market Getting Tighter," Hydrocarbon Processing,
     pp. 201-204, Nov. 1973.

88. "Oil Crunch Spurs Waste Oil Reclaiming," C&E News, vol. 52, p. 26,
     Feb. 1974.

89. "Re-refining Lures Oil Firms," C&E News, vol. 82, no. 7, p. 64, Mar.
     1975.

90. Sanner, W. S.; Ortuglio, C.; Walters, J. G.; and Wolfson, D. E. Conversion of Municipal and Industrial Refuse into Useful Materials by Pyrolysis, U. S. Bureau of Mines Report of Investigations 7428, 1970.

91. Cheremisinoff, P. N. and Rao, K. B. "Fluid Bed Reactors," Pollution Engineering, vol. 4, no. 6, 1972.

92. Coe, W. B. Turk, M. Symposium: Processing Agricultural and Municipal Wastes, The Avi Publishing Co., Inc., 1973, pp. 29-37.

93. Loehr, R. C. "Animal Wastes: A National Problem," Journal of the Sanitary Engineering Division (ASCE) Proceedings, vol. 85, pp. 189-221, April 1969.

94. Cummings, F. R. Private correspondence, 1975.

95. Humphrey, A. E. "Current Developments in Fermentation," Chemical Engineering, Dec. 9, 1974, pp. 98-112.

96. Hart, S. A. Turner, M. E. "Lagoons for Livestock Manure," Journal WPCF, vol. 37, no. 11, Nov. 1965, pp. 1578-1596.

97. Commentator, "Cattle Manure Will Be Biogasified to Methane," C&E News, Oct. 1965.

98. McKinney, R. E. Microbiology for Sanitary Engineers, McGraw-Hill, New York, 1962, pp. 249-256.

99. Kufel, J. M. Generation of Methane in Sanitary Landfills, Master's project, New Jersey Institute of Technology, 1975.

100. Development of Construction and Use Criteria for Sanitary Landfills, An interim report, prepared by County of Los Angeles, Department of County Engineer, and Engineering Science, Inc., Arcadia, Calif., under Grant #DOT-UI-00046 for U. S. Department of Health, Education, and Welfare, 1969.

101. Sorg, T. J. Hickman, H. L. Sanitary Landfill, Facts, Report SW-4ts, U. S. Department of Health, Education, and Welfare, 1970.

102. Brunner, D. R. and Keller, D. J. Sanitary Landfill Design and Operation, Report SW-65ts, U. S. Environmental Protection Agency, 1972.

103. Burchinal, J. C. Microbiology and Acid Production in Sanitary Landfills, An interim report, supported by the U. S. Department of Health, Education, and Welfare, under Grant #UI00520-02, 1967.

104. Hagerty, D. J.; Pavoni, J. L.; and Heer, J. E. Solid Waste Management, Van Nostrand Reinhold, New York, 1973.

105. Rhyne, C. W. Landfill Gas, paper, Office of Solid Waste Management
      Programs, USEPA, April 4, 1974.

106. Farquhar, G. J. and Rovers, F. A. Gas Production during Refuse
      Decomposition, paper, Civil Engineering, University of Waterloo,
      Waterloo, Ontario.

107. Jeris, J. S. and McCarty, P. L. "The Biochemistry of Methane
      Fermentation Using C^{14} Tracers," Journal WPCF, vol. 37, no. 2,
      Feb. 1965.

108. Zajic, J. E. Water Pollution, Marcel Dekker, N. Y., vol. 1, 1971.

109. Burdon, K. L. and Williams, R. P. Microbiology, MacMillan, New York,
      6th Ed., 1968.

110. Barker, H. A. "Biological Formation of Methane," Industrial and
      Engineering Chemistry, vol. 48, no. 9, Sept. 1956.

111. Pelczar, M. J. and Reid, R. D. Microbiology, McGraw-Hill, New York,
      1972.

112. Lawrence, A. W. and McCarty, P. L. "The Role of Sulfide in Preventing
      Heavy Metal Toxicity in Anaerobic Treatment," Journal WPCF, vol. 37,
      no. 113, March 1965.

113. Metcalf and Eddy, Wastewater Engineering, McGraw-Hill, New York, 1972.

114. Sawyer, C. N. and McCarthy, P. L. Chemistry for Sanitary Engineers,
      2nd ed., McGraw-Hill, New York, 1967.

115. Operation of Wastewater Treatment Plants, WPCF Manual of Practice
      No. 11, WPCF, Washington, D. C. 1970.

116. Sanders, F. A. and Bloodgood, D. E. "The Effect of Nitrogen to
      Carbon Ratio on Anaerobic Decomposition," Journal WPCF, vol. 37,
      no. 12, Dec. 1965.

117. Kugelman, I. J. and McCarty, P. L. "Cation Toxicity and Stimulation
      in Anaerobic Waste Treatment," Journal WPCF, vol. 37, no. 1, Jan. 1965.

118. Bishop, W. D.; Cater, R. C.; and Ludwig, H. F. "Gas Movement In Land-
      filled Rubbish," Public Works, Nov. 1965.

119. Merz, R. C. and Stone, R. Special Studies of a Sanitary Landfill,
      Grant #UI00518-08, Final Summary Report, Jan. 1964 to Dec. 1968,
      to U. S. Department of Health, Education, and Welfare.

120. McKinney, R. E. Microbiology for Sanitary Engineers, McGraw-Hill,
      New York, 1962.

121. Merz, R. C. and Stone, R. "Gas Production in a Sanitary Landfill," Public Works, Feb. 1964.

122. Eliassen, Rolf "Decomposition of Landfills," American Journal of Public Health, vol. 32, Sept. 1942.

123. Sonoma County Solid Waste Stabilization Study, Emcon Assoc. to U. S. EPA, Grant #G06-EC-00351, EPA-SW-530-65d, 1974.

124. Reinhardt, J. J. and Ham, R. K. Solid Waste Milling and Disposal on Land Without Cover, EPA report #530/SW-62d.1, 1966-72.

125. Carlson, J. A. Shoreline Regional Park Gas Recovery Program, EPA Grant #S803396 01, Progress report for Phase 1, July 1, 1974 to Jan. 31, 1975.

126. Granthan, J. B.; Estep, E. M.; Pierovich, J. M.; Tarkow, H.; and Adams, T. C. Energy and Raw Material Potentials of Wood Residue in the Pacific Coast States, Forest Service Portland, Oregon, USDA, Pacific Northwest Forest and Range Experiment Station, 1973.

127. Wood Waste Utilization: An Overview of a Workshop University of Tennessee Environment Center, Discussion draft, Knoxville, Tenn., May, 1975.

128. Wood Waste Turns on the Heat, Orr & Sembower, Inc., Middletown, Penn., Published information.

129. Brown, Owen. "Steam Generation with Bark." In Converting Bark into Opportunities (ed. by A. C. Van Vliet). Proceedings of a Conference, Oregon State University, School of Forestry, Corvallis, pp. 43-45, Nov. 1971.

130. Corder, S. E.; Scroggins, T. L.; Meade, W. E.; and Everson, G. D. Wood and Bark Residues in Oregon: Trends in Their Use, Oregon State University, Forest Research Laboratory, Corvallis, Research Paper 11, 16 p. March, 1972.

131. "Wood Waste Fed Boiler Systems," Wood & Wood Products, July, 1974.

132. Solid Waste Converter, York-Shipley, Inc., York, Penn., Published Information.

133. Koch, Peter and Mullen, J. F. "Bark from Southern Pine May Find Use As Fuel," Forest Industries, vol. 98, no. 4, pp. 36-37, April, 1971.

134. Haley, Thomas I. "Hemlock Bark as Fuel for Steam Generation," In Converting Bark into Opportunities (ed. by A. C. Van Vliet). Proceedings of a Conference, Oregon State University, School of Forestry, Corvallis, pp. 51-52, Nov. 1971.

135. Pratt, Willard E. Experience with an Automatic Water-Wood Fired Steam Plant, paper presented at a meeting of the Northern California Section of the Forest Products Research Soc., Fresno, Calif., pp. 8-13, April 20, 1967.

136. Converting Leaves to Logs Makes Unusual Business, Norman Malone Associates, Inc., Akron, Ohio, News release, 1975.

137. Sludge Treatment and Disposal, U. S. Environmental Protection Agency Technology Transfer, Oct. 1974.

138. Fair, G. M. and Geyer, J. C. Water Supply and Wastewater Disposal, John Wiley, New York, 1958.

139. "Anaerobic Sludge Digestion," JWPCF, MOP, no. 16, 1968.

140. Burd, R. S. A Study of Sludge Handling and Disposal, Federal Control Administration, Publication WP-20-4, May, 1968.

141. Schroepfer, G. J. and Ziemke, N. R. "Development of the Anaerobic Contact Process: I. Pilot Plant Investigations and Economics," Sewage and Industrial Wastes, vol. 31, no. 2, pp. 164-190.

142. McCarty, P. L. "Kinetics of Waste Assimilation in Anaerobic Treatment," In Developments in Industrial Microbiology, American Institute of Biological Sciences, vol. 7, 1966.

143. Ritter, E. L. "Design and Operating Experiences Using Diffused Aeration of Sludge Digestion," JWPCF, vol. 42, no. 10, p. 1782, 1970.

144. Process Design Manual for Upgrading Existing Wastewater Treatment Plants, Technology Transfer, U. S. Environmental Protection Agency, Washington, D. C., Oct. 1974.

145. McCarty, P. L. "Anaerobic Waste Treatment Fundamentals," Public Works, vol. 95, pp. 107-112, 1964.

146. Sawyer, C. N. and McCarty, P. L. Chemistry for Sanitary Engineers, McGraw-Hill, New York, 1967.

147. Speece, R. L. and McCarty, P. L. "Nutrient Requirements and Biological Solids Accumulation in Anaerobic Digestion," In Proceedings of the International Conference on Water Pollution Resources, Pergemon Press, New York, 1962.

148. McCarty, P. L. "Anaerobic Waste Treatment Fundamentals, Part 3, Toxic Materials and Their Control," Public Works, pp. 91-94, Nov. 1964.

149. Kugelman, I. J. and Chin, K. K. Toxicity Synergism and Antagonism in Anaerobic Waste Treatment Processes. Presented before Division of Air, Water and Waste Chemistry, ACS, Houston, Feb. 1970.

150. Barth, E. F. "Summary Report on the Effects of Heavy Metals on the Biological Treatment Processes," JWPCF, vol. 37, no. 1, pp. 86-96, 1965.

151. Lawrence, A. W. and McCarty, P. L. "The Role of Sulfide in Preventing Heavy Metal Toxicity in Anaerobic Treatment," JWPCF, vol. 37, pp. 392-406, 1965.

152. Kugelman, I. J. and McCarty, P. L. "Cation Toxicity and Stimulation in Anaerobic Waste Treatment," JWPCF.

153. McCarty, P. L. and Brosseau, M. H. Toxic Effect of Individual Volatile Acids in Anaerobic Treatment. Proceedings 18th Industrial Wastes Conference, Purdue University, 1963.

154. Lawrence, A. W.; Kugelman, I. J.; and McCarty, P. L. "Ion Effects in Aerobic Digestion," Technical Report No. 33, Department of Civil Engineering, Stanford University, March 1964.

155. Weland, A. F. and Cheremisinoff, P. N. "Anaerobic Digestion of Sludge," Water & Sewage Works, vol. 122, nos. 10, 11, 1975.

156. Bryan, A. C. and Ganett, M. T., Jr. "What Do You Do With Sludge? Houston Has The Answer," Public Works, Dec. 1972.

157. Hill, G. "New Sewage Plan Studied on Coast," New York Times, Oct. 5, 1975, p. 34.

158. Committee on Refuse Disposal, American Public Works Association, Municipal Refuse Disposal, Interstate Printers and Publishers, Inc., Danville, Ill., 1961.

159. Engdahl, Richard B. Solid Waste Processing, U. S. Government Printing Office, Washington, D. C., 1969.

160. Floros, James. "Designing for Scrap Stool Magnetic Removal," Pollution Engineering, vol. 4, no. 7, Oct. 1972, pp. 32-33.

161. Yen, T. F. ed. Recycling and Disposal of Solid Wastes, Ann Arbor Science Publishers, Inc., Ann Arbor, Mich., 1974.

162. Bond, Richard G. and Conrad, P. Straub, eds. Handbook of Environmental Control, vol. 2, The Chemical Rubber Co., 1973.

163. Mallan, G. M. and Titlow, E. I. Energy and Resource Recovery from Solid Wastes. Presented to Washington Academy of Sciences University of Maryland, College Park, Md., March 13, 14, 1975.

164. Salvato, Joseph A., Jr. Environmental Engineering and Sanitation, Wiley Interscience, New York, 1972.

165. Resource Recovery from Municipal Solid Wastes, National Center for
      Resource Recovery, Inc., Lexington Books, D. C., Heath & Co.,
      Lexington, Mass., 1974.

166. WHPA "Disposal of Solid Wastes," Chemical Engineering, vol. 78,
      no. 22, 1971.

167. Cannon, J. "Steel: The Recyclable Material," Environment, vol. 15,
      no. 9, Nov. 1973.

INDEX